Seismic Performance of Asymmetric Building Structures

Seismic Performance of Asymmetric Building Structures

Chunwei Zhang

School of Civil Engineering, Qingdao University of Technology, China

Zeshan Alam

Centre for Infrastructure Engineering, Western Sydney University, Australia
School of Civil Engineering, Qingdao University of Technology, China

Li Sun

School of Civil Engineering, Shenyang Jianzhu University, China

Bijan Samali

Centre for Infrastructure Engineering, Western Sydney University, Australia
School of Civil Engineering, Qingdao University of Technology, China

CRC Press
Taylor & Francis Group
Boca Raton London New York Leiden

CRC Press is an imprint of the
Taylor & Francis Group, an **informa** business

A BALKEMA BOOK

CRC Press/Balkema is an imprint of the Taylor & Francis Group, an informa business

© 2020 Taylor & Francis Group, London, UK

Typeset by Apex CoVantage, LLC

Library of Congress Cataloging-in-Publication Data
Names: Zhang, Chunwei, (Civil Engineer), author. | Alam, Zeshan, author. | Sun, Li, author. |
    Samali, Bijan, author.
Title: Seismic performance of asymmetric building structures / by Chunwei Zhang,
    Zeshan Alam, Li Sun and Bijan Samali.
Description: Boca Raton : CRC Press, 2020.
Identifiers: LCCN 2020003005 (print) | LCCN 2020003006 (ebook) | ISBN 9780367903435 (hbk) |
    ISBN 9781003026556 (ebook)
Subjects: LCSH: Earthquake resistant design—Evaluation.
Classification: LCC TA658.44 .Z43 2020 (print) | LCC TA658.44 (ebook) | DDC 693.8/52—dc23
LC record available at https://lccn.loc.gov/2020003005
LC ebook record available at https://lccn.loc.gov/2020003006

Published by: CRC Press/Balkema
                Schipholweg 107C, 2316 XC Leiden, The Netherlands
                e-mail: Pub.NL@taylorandfrancis.com
                www.crcpress.com – www.taylorandfrancis.com

ISBN: 978-0-367-90343-5 (Hbk)
ISBN: 978-1-003-02655-6 (eBook)
DOI: https://doi.org/10.1201/9781003026556

Contents

Figures

Tables

Acknowledgments

The book is financially supported by the Ministry of Science and Technology of China (Grant No. 2017YFC0703603), the National Natural Science Foundation of China (Grant No. 51678322), the Taishan Scholar Priority Discipline Talent Group program funded by Shandong Province, the Cooperative Innovation Center of Engineering Construction and Safety in the Shandong Blue Economic Zone and the first-class discipline project funded by the Education Department of Shandong Province.

Abbreviations

3D	Three dimensional
C-1	Concrete model
DBE	Design basis earthquake
ELF	Equivalent lateral force
Exp.	Experimental
FBG	Fiber Bragg grating
FE	Finite element
FS	Flexible side
IDR	Inter-story drift ratio
IRI	Irregularly irregular
KI	Stiffness irregularity
LFRS	Lateral force resisting system
MCE	Maximum considered earthquake
MDOF	Multi degrees of freedom
MI	Mass irregularity
PGA	Peak ground acceleration
RC	Reinforced concrete
RI	Regularly irregular
S-1	Steel model no. 1
S-2	Steel model no. 2
S-3	Steel model no. 3
S-4	Steel model no. 4
SDC	Seismic design category
SDOF	Single degree of freedom
SI	Soft story mechanism
SS	Stiff side
TB	Torsionally balanced
TF	Torsionally flexible
TF-IRI	Torsionally flexible-irregularly irregular
TF-RI	Torsionally flexible-regularly irregular
TS	Torsionally stiff
TS-IRI	Torsionally stiff-irregularly irregular
TS-RI	Torsionally stiff-regularly irregular

TU   Torsionally unbalanced
TU-TF   Torsionally unbalanced-torsionally flexible
TU-TS   Torsionally unbalanced-torsionally stiff
VI   Strength irregularity
WI   Weak story mechanism

Principal notations

Δ_{li}	Displacement vector in the longitudinal direction
Δ_{ti}	Displacement vector in the transverse direction
$\Delta_{\theta i}$	Displacement vector in the vertical direction
$J'_{\varnothing,G}$	Global polar moment of inertia
$K_{\varnothing}$	Rotational stiffness
k_{li} and k_{ti}	Directional lateral stiffness matrices of the orthogonal plane
K_x	Lateral stiffness in X-direction
K_y	Lateral stiffness in Y-direction
$k_{\theta i}$	Torsional stiffness matrix in the local coordinates
$\ddot{u}_g$	Ground motion along longitudinal direction
$\ddot{v}_g$	Ground motion along transverse direction
α_T	Temperature sensitivity coefficient of the FBG sensor
α_{xG}	Global mass center coordinate in the X-direction
α_{yG}	Global mass center coordinate in the Y-direction
α_c	Strain sensitivity coefficient of the FBG sensor
$\beta x, \ CR$	Stiffness center coordinate along X-direction
$\beta x, \ CR$	Stiffness center coordinate along Y-direction
δ_{li}	Shear vector in the longitudinal direction
δ_{ti}	Shear vector in the transverse direction
$\delta_{\theta i}$	Shear vector in the vertical direction
ε_C	Compressive strain
ε_C^{in}	Inelastic compressive strain
$\varepsilon_{ck.}^{ten.}$	Tensile cracking strain
ε_{cr}	Cracking strain
ε_E	Engineering strain
$\varepsilon_{el.}^{ten.}$	Elastic tensile strain
ε_{el}^c	Elastic compressive strain
$\varepsilon_{tot.}^{ten.}$	Total tensile strain
η_{eff}	Effective refractive index
λ_B	Center wavelength of FBG

λ_p	Center wavelength of FBG during pouring
λ_t	Center wavelength of FBG during test
ξ_k	First vibration damping ratio
ξ_l	Third vibration damping ratio
σ_C	Compressive stress
σ_{cu}	Ultimate compressive stress
σ_E	Engineering stress
σ_T	True stress
$\ddot{\varphi}_{gz}$	Vertical ground motion component
ω_k	First vibration frequency
ω_l	Third vibration frequency
$\omega_{x,\,G}$	Global uncoupled frequency in X-direction
$\omega_{y,\,G}$	Global uncoupled frequency in Y-direction
A_X	Torsional irregularity coefficient
e_a	Accidental design eccentricity
e_d^1	Primary design eccentricity
e_d^2	Secondary design eccentricity
$\varnothing$	Arbitrary seismic orientation
ΔT	Change in the temperature
C	Global damping matrix
C_M	Center of mass
C_R	Center of stiffness
C_V	Center of strength
d_c	Compressive damage parameter
E	Elastic modulus
$e/e_s/e_m/e_v$	Eccentricity/stiffness eccentricity/mass eccentricity/strength eccentricity
e_{mx}	Mass eccentricity in the X-direction
e_{my}	Mass eccentricity in the Y-direction
e_{sx}	Stiffness eccentricity in the X-direction
e_{sy}	Stiffness eccentricity in the Y-direction
e_{vx}	Strength eccentricity in the X-direction
e_{vy}	Strength eccentricity in the Y-direction
g	Acceleration of gravity
K	Global stiffness matrix
K_c	Ratio of the second stress invariant on the tensile meridian to compressive meridian
M	Global mass matrix
N	Similarity factor
T_p	Pouring temperature
T_t	Test temperature
u_x	Displacement vector in the X-direction

u_y	Displacement vector in the Y-direction
u_θ	Rotational displacement vector
β	Percentage coefficient for accidental eccentricity
ε	Strain
ε_0	Strain at fc'
Θ	Response quantity
Λ	FBG period
σ	Stress
θ	Angular drift
ψ	Dilation angle
ψ	Degree of freedom
Ω	Uncoupled torsional frequency ratio
f_{b0}/f_{c0}	Ratio of the compressive strength under biaxial loading to uniaxial compressive strength
fc'	Compressive strength of concrete
α	Amplification coefficient for the flexible edge
δ	Amplification coefficient for the stiff edge
υ_c	Poisson's ratio

Chapter 1

Introduction to the seismic performance of asymmetric building structures

1.1 Introduction and background

Seismic events stand out amongst nature's most prominent threats to civil infrastructure facilities and human life because throughout the years they have caused the devastation of urban areas and cities on almost every continent prone to seismic risk. They are the least comprehended among natural hazards, and the most critical distinction of such hazards is that the risk to life is associated with man-made structures. Seismic hazard poses a unique engineering design problem, and the optimum engineering approach is to design the structure so as to avoid a collapse under major earthquakes. This ensures no loss of life and accepts the possibility of damage under rare earthquakes. With exception to the landslides under severe earthquakes, seismic impacts that lead to broad death tolls are collapses of buildings, bridges, dams and other man-made civil infrastructure facilities. To ensure the safety of these structures, only the development of seismic-resistant design strategies can counter the seismic risk associated with these structures. This demands successful implementation of engineering knowledge to avoid damage to the building structures under seismic actions.

From previous research investigations corresponding to the collapse of building structures under earthquakes, it has been well understood that one of the major reasons of the collapse of these buildings was the irregularity of these structures (Sharma *et al.*, 2016; Bikçe and Çelik, 2016; Shakya and Kawan, 2016; Zhao *et al.*, 2009; Varum *et al.*, 2018). Irregularity is a term associated to building structures when their centers of mass (C_M) and stiffness (C_R) are no-coincident (Hejal and Chopra, 1989b; Harasimowicz and Goel, 1998; Chandler and Hutchinson, 1986). Such structures exhibit higher probability to get damaged under seismic actions compared with counter regular structures. With recent developments in infrastructure, the advancement in the architecture of building structures has significantly improved (Fig. 1.1). However, irregularities tend to develop when aesthetic/functional considerations are made during the architectural planning of these structures. Torsional response can occur for various reasons even in symmetric structures that are highly favored in current earthquake resistant design provisions of various codes. These reasons include but are not limited to rotational components of seismic excitation, non-uniform yielding in the structure, random distribution of the internal loads, inelastic behavior and site-specific characteristics of the seismic excitation. Hence, to some degree, almost all real-life structures experience torsional vibrations. Therefore, perfect symmetry is an idealization and very rarely occurs as real-life structures are almost always asymmetric. Irregularities in a structure are very difficult to define as they dramatically vary in their nature and in principle. Broadly classifying the structural irregularities, they can be of two types: (1) plan-asymmetric

Figure 1.1 Silicon Alley, New York representing different asymmetric structures (Photo credit: Dave Lindblom, https://upload.wikimedia.org/wikipedia/commons/6/67/Flatiron_District.jpg)

structures and (2) vertically asymmetric structures. Starting with plan-asymmetry, previously executed studies demonstrate that this type of irregularity results in severe damage to the plan-asymmetric structures since it produces floor rotations along with floor translations (torsional coupling). Such irregularities form because of the irregular distributions of stiffness/strength and mass in the structure. Various investigations in the past have evaluated the impacts of torsional coupling in plan-asymmetric structures by considering simplified one-story analytical models. Such models were considered for the development of design provisions for various one-story asymmetric structures as well as multi-story asymmetric structures. However, these models are not justified for multi-story asymmetric structures especially for the inelastic response and therefore, such numerical investigations are only applicable to a few cases of the realistic asymmetric structures. Despite the fact that these models are incapable of addressing the problem of asymmetry very well, they are still used by numerous researchers to provide particular qualitative demonstration of the problem.

The components of a moment resisting structure, resisting the seismic actions, are defined as lateral force resisting systems (LFRS). In asymmetric structures, damage is generally formed by stress concentration at weak locations of the LRFS. These weak locations cause further deterioration of the structural components thereby causing the structural failure. The reason behind the stress concentration at weak locations is the inherent eccentricity (difference of C_M and C_R) in the structural floor plan. Asymmetric structures can be broadly classified as plan-asymmetric and vertically asymmetric. A typical example of plan-asymmetry (De Stefano *et al.*, 1998; Ghersi and Rossi, 2001; Bhatt and Bento, 2014) is illustrated in Figure 1.2.

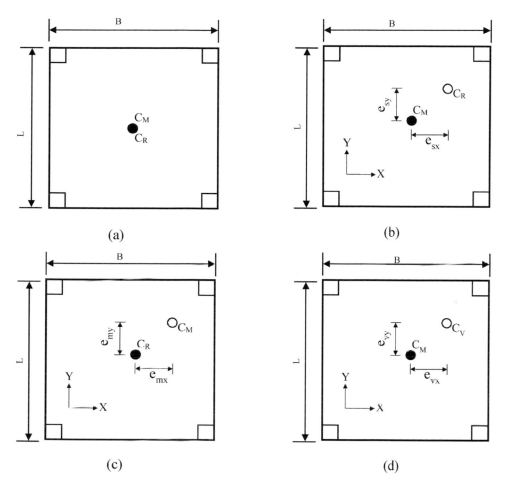

Figure 1.2 Typical example of symmetric and plan-asymmetric structures: (a) symmetric system, (b) stiffness eccentric system, (c) mass eccentric system and (d) strength eccentric system

Similarly, vertical-asymmetry can be defined as the irregular distribution of stiffness/strength and/or mass along the structure's height. There are various reasons for real-life structures to be vertically asymmetric. For instance, in majority of the commercial structures, basements are created by eliminating central columns which eventually leads to the soft story mechanism at basement level. Another practical reason behind vertical-asymmetry is the reduction in the sizes of columns and beams in the higher order floors mainly for two purposes: (1) to fulfill functional requirements such as storing heavy mechanical appliances and (2) cost-reduction of the construction. This creates difference in the vertical distribution of stiffness/strength and/or mass with respect to the adjacent floors. Besides, there are several other reasons for the formation of vertical-asymmetry such as variation in the material properties, improper construction methods and construction mistakes. Typical examples of vertically regular and irregular structures (Moehle, 1984; Truman and Cheng, 1990; Das and Nau, 2003; Nezhad and Poursha, 2015; Basu and Giri, 2015) are illustrated in Figure 1.3.

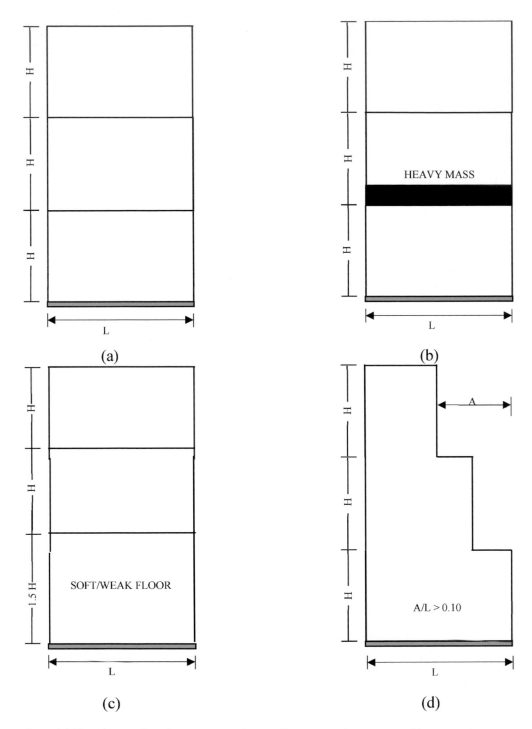

Figure 1.3 Typical examples of symmetric and vertically asymmetric structures: (a) symmetric system, (b) irregular mass distribution, (c) soft story structure and (d) setback structure

This chapter provides a detailed picture of the work presented in previous studies on the damage response of vertical and plan-asymmetric structures under the influence of torsional vibration, mainly highlighting the research background, challenges addressed in this book, objectives of the book, methodology adopted to address the challenges and finally an outline of the book.

1.2   Challenges addressed in this book

Post-earthquake inspections have indicated stress concentration at weaker locations as the main cause of damage to asymmetric structures. In this regard, seismic design codes acknowledge the relevance of torsional coupling to structural damage response. To estimate the contribution of torsion on the seismic response, several simplified procedures have been established by many researchers. In general, two kinds of eccentricities are considered in the seismic design: (1) static eccentricity (natural or inherent eccentricity) and (2) dynamic eccentricity (multiple of accidental eccentricity and a coefficient). The common design practice is to use accidental torsion to indirectly account for (a) uneven yielding of structural members, (b) changes in the geometric and material properties of the LRFS, (c) difference in the distribution of designed mass and reactive mass and (d) varying orientations of seismic excitations. The potential damage to the buildings from previous major earthquakes is evident, and the main cause of structural damage is the local stress concentration at weak locations in the buildings. Asymmetric structures are more vulnerable to damage at these weak locations than their regular counterparts. The current design practice to deal with the potential influence of torsional vibrations in the asymmetric structures has been established on over-simplified procedures, and in relevance to this issue, despite the compliance of asymmetric structures with the seismic design guidelines, the asymmetric structures are still potentially vulnerable to damage.

From previous research studies and the provisions of various international design standards, it is evident that the problem of structural irregularity has been viewed as a global parameter. However, location-specific influence corresponding to the seismic damage response has not been explored very well. Moreover, in recent times, development in modern architecture has evolved which has led to the presence of more severe circumstances of irregularities in the structure. For instance, a structure can have multiple irregularities in its plan and elevation at the same time. The interaction of these irregularities may lead to more severe damage. The current design practice considers these irregularities separately. Moreover, the whole emphasis is only on the global response, ignoring entirely the local distress in the structure. Therefore, the primary emphasis of this book is to evaluate the damage characteristics of asymmetric structures keeping in view both the local and global response perspective under the interaction of irregularities. The research target is achieved by evaluating the damage behavior of plan-asymmetric RC structure. The plan-asymmetric RC structure was initially tested in elastic range, and then the structure was transformed into a highly inelastic state by progressively increasing the seismic excitation. However, to simulate the damage-based results for a wide range of asymmetric scenarios, an extensive experimental investigation on vertical and plan-asymmetric steel structures has been carried out. All the experimental models were carefully equipped with calibrated instruments to thoroughly investigate both local and global structural response.

Through the observations of the structural failure phenomenon in the simulated seismic shake table tests, the damage mechanism is analyzed, and the corresponding observations have been established to study the weak links in the structure and establish seismic design guidelines. Moreover, for monitoring and inspection of existing asymmetric structures, it is a huge challenge to determine the structure's state of collapse and the location where it is likely to occur. Using the findings of this book, a comprehensive guideline is established to determine the state of possible threat to asymmetric structures under seismic actions.

1.3   Objectives of this book

The main focus of this research is to produce, by both experimental and numerical investigations, a reliable connection between the seismic damage response and the influence of torsional vibrations in vertical and plan-asymmetric structures. The purpose of establishing this connection is to provide detailed information about critical locations in asymmetric structures and the development of improved seismic design guidelines. In this regard, following are the objectives of this research:

- To carry out a feasibility study on applications of FBG strain sensors for effective monitoring of damage response in the asymmetric structures under dynamic loading. The feasibility and successful implementation of the FBG sensors were ensured by comparing the damage response with varying dynamic characteristics.
- To conduct detailed experimental testing on plan-asymmetric RC structure under progressive seismic excitation and to monitor the local damage and global response during the transformation of the structure from elastic to inelastic state.
- To investigate the behavior of local seismic response at the FS of the structure experiencing tensile/compressive deformations.
- To establish the parameter for the prediction of the cracks and formation of plastic hinges at weak locations.
- To construct a wide range of asymmetric steel models ranging from highly torsionally stiff (TS) to highly torsionally flexible (TF) and then conduct detailed experimental testing for local/global response simulation.
- To develop the strain contour plots for the assessment of the local response transition from the FS to the SS of the structure.
- To determine the parameters influencing the local/global response under torsional vibrations.
- To develop numerical models for local/global response validation.
  - To assess the complex behavior of local response at the FS of the asymmetric structure and to evaluate the uncertainties associated with global response using validated numerical models.
- To propose seismic design guidelines for asymmetric structures based on the cumulative knowledge established through experimental and numerical investigations.

1.4   Methodology adopted to address the described challenges

To achieve the objectives, the research program was divided into four groups: Detailed experimental work on the concrete model, detailed experimental investigations on asymmetric steel models under four different seismic excitations, development of the FE model to verify the complex and abnormal behavior of local seismic response, and numerical studies using finite

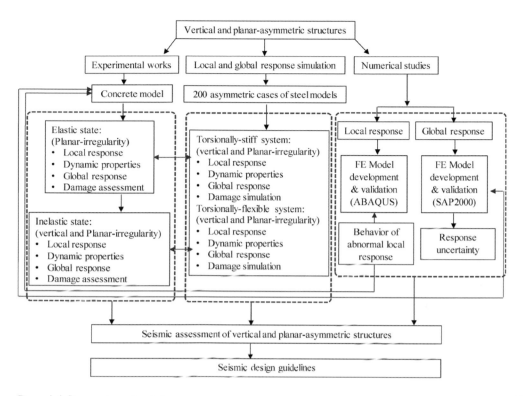

Figure 1.4 Research methodology

element (FE) analysis to quantify the global response uncertainty. A summary of the methodology adopted in this book to achieve the described objectives is presented in Figure 1.4.

1.4.1   Experimental work

In the experimental research work, first, a plan-asymmetric reinforced concrete (RC) model was tested for both local and global seismic response. Based on the structure's transformation from the elastic state into an inelastic state, various parameters corresponding to each damage state of the structure were investigated considering the influencing of asymmetry of the structure. For a wide range of asymmetric structures, damage response was simulated using steel models. The following physical models are considered in this research:

- C-1 model: Three-story reinforced concrete (RC) frame-shear wall structure with mono-directional stiffness eccentricities in the floor plan. This model is a torsionally unbalanced-torsionally stiff (TU-TS) model in the elastic state and torsionally unbalanced-torsionally flexible (TU-TF) in the inelastic state.
- S-1 model: This category of structures corresponds to TU-TS models with bi-eccentric and mono-eccentric square-shape structures having stiffness eccentricity at all floor levels in the reference state. The term reference state is defined in Chapter 3. There are two structures in this category: bi-directional asymmetric and mono-symmetric.

- S-2 model: This category of structures corresponds to TU-TS models with bi-eccentric square-shape structures having stiffness and strength eccentricities.
- S-3 model: This category of structures corresponds to TU-TS models with bi-eccentric L-shape structures having stiffness eccentricities at all floor levels.
- S-4 model: This category of structures corresponds to TU-TF models with bi-eccentric circular-shape structures having stiffness and strength eccentricities at all floor levels.

Steel models were constructed with their regular counterparts to better demonstrate the enhancement in seismic demands under the influence of coupled torsional vibrations. Models corresponding to each type were then used to develop further asymmetric structures as described in Chapter 3 by shifting the C_M at each floor separately or at all floors simultaneously, thereby developing 25 subcases for each model.

1.4.2   Numerical investigation

Numerical studies were conducted by establishing a validated FE model in ABAQUS to investigate the complex behavior of local response at the FS of the structure. For global response investigation and to observe the uncertainties associated with global response, FE models in SAP2000 were developed and validated with the global response of the plan-asymmetric RC structure.

1.5   Outline of the work

To achieve the objectives of this research, several tasks and sub-tasks were identified where each task refers to each chapter of the book. This book is organized into eight chapters after this one. A short detail of each chapter of this book is as follows:

Chapter 2 presents background information for vertical and plan-asymmetric structures. This chapter includes a brief description of planar and vertical irregularities and their sub-types. Previous numerical studies on single-story and multi-story asymmetric structures are summarized to provide a general understanding of the behavior of asymmetric structures. Experimental data stock on asymmetric structures is very limited. However, a summary of previous experimental studies on asymmetric structures is provided in this chapter. Investigations concerning the major damage to asymmetric structures during the past earthquakes are summarized to indicate the potential threat to asymmetric structures against seismic excitations. Several shake table studies for damage evaluation are summarized; however, the existing data stock is very limited for research on local response investigation in asymmetric structures. Finally, a summary is presented on the basis of a thorough literature review, providing the research gap and connection of this research with previous investigations.

Chapter 3 presents the experimental strategy, geometric and material properties of model, instrumentation and characteristics of the seismic excitations used during the experiments. In the experimental strategies, an explanation about the two types of models is presented. One is plan-asymmetric structure, and the other corresponds to vertical and plan-asymmetric steel models. In the section on materials and geometric details, material characteristics, section sizes and similitude law are presented. In instrumentation planning, details about the instruments used during the experiments and their strategic locations are presented. Finally, the characteristics of the seismic excitations are demonstrated in Chapter 7.

Chapter 4 discusses the experimental findings corresponding to the damage behavior of plan-asymmetric RC structure and simulated damage behavior for steel models. Influence of the local seismic response is explained considering the asymmetric properties of the structures.

Chapter 5 describes the development of a validated FE model in ABAQUS and demonstrates the complex nature of local seismic response at the FS of the RC plan-asymmetric structure. The abnormal behavior of the local response is explained as the consequent influence of eccentricities in the structure.

Chapter 6 discusses the experimental results corresponding to the global behavior of plan-asymmetric RC structure in the elastic and inelastic state and simulated global behavior for steel models. Influence of the global seismic response is explained considering the asymmetric properties of the structures.

Chapter 7 explains the development of a validated FE model in SAP2000 for the evaluation of uncertainties associated with the global response and demonstrates the influence of varying global response under varying characteristics of seismic excitations.

Chapter 8 explains the design guidelines established on the basis of experimental and numerical findings. This chapter also discusses the potential drawbacks in the current seismic codes in regards to the seismic design of asymmetric structures.

Chapter 9 summarizes the conclusions drawn from extensive research presented in this book.

Chapter 2

A review of research on design guidelines and seismic performance of asymmetric building structures

2.1   Introduction

Structural vulnerability to damage under earthquakes has remained a key area for researchers over the past few decades in order to minimize the earthquake hazard to building structures as much as possible. Past studies pertaining to the structural damage under earthquakes indicate structural asymmetry as one of the most influencing parameters causing damage to the structures. This eventually leads to the need for better understanding of structures possessing planar and/or vertical irregularities. With the recent advancement in civil infrastructure facilities, modern structures tend to possess obvious irregularities, which eventually leads them towards damage under seismic excitations as the current seismic design provisions are not sufficient to fully address the potential consequences of structural asymmetry. Moreover, these seismic design provisions are based on simplified single-story analytical models and are not representative of multi-story structures. The reason behind the insufficiency of simplified single-story models as representative multi-story models is the complex nature of seismic response in multi-story asymmetric structures. Besides, multi-story structures are likely to possess higher mode effects whereas the considered models are not representative of such cases. In this regard, it is imperative to carry out detailed investigation on multi-story asymmetric structures under the interaction of various irregularities, both experimentally and numerically.

Several studies have aimed at drawing some definitive conclusions on the behavior of asymmetric structures. One such study was conducted by Peruš and Fajfar (2005), where an issue of a general nature was investigated using parametric analysis. The research aimed at evaluating the effects of elastic and plastic excursions on torsional response. The major findings of the parametric analysis indicated that global behavior of the asymmetric structures in the elastic and inelastic states is similar. The research established this conclusion based on the fact that elastic and inelastic responses are more amplified in the translational engineering demand, rather than the rotational engineering demands. Numerous researchers have contributed towards the investigations on the influence of torsion in asymmetric structures. Some of the pioneer works on the subject matter have been reported in the literature (Kan and Chopra, 1981a, 1981b; Dempsey and Tso, 1982; Hejal and Chopra, 1989a, 1989b; Goel and Chopra, 1990; Chopra and Goel, 1991; Tso and Zhu, 1992; Zhu and Tso, 1992; Humar and Kumar, 1998a, 1998b; Dutta and Das, 2002a, 2002b; Moehle and Alarcon, 1986).

2.2   Research on single-story asymmetric structures

As has been discussed in the previous chapter, earlier investigations on asymmetric structures are only limited to simplified single-story models and the primary reason for considering such model is its simplicity and ease with the calculation process. The current design practice utilizes the formulations that were developed using such simplified models. However, with the advancement in various computational tools, various investigations have been conducted on multi-story asymmetric structures. These investigations have demonstrated the complicated nature of seismic response and therefore, simplified one-story models are still favored by numerous researchers (Ladinovic, 2008; Lignos and Gantes, 2005; Lucchini *et al.*, 2010). In terms of research investigations on one-story structures, the majority of the previous studies are based on the variation of positions of C_R (center of stiffness) and/or C_V (center of strength) and C_M (center of mass) with respect to each as the difference in C_R and C_M eventually leads to the formation of floor eccentricity thereby causing torsional coupling in the seismic response. These floor eccentricities were then investigated by varying the location of C_R while maintaining a constant location for C_M; similarly, these eccentricities were also investigated by varying the location of C_M while maintaining a constant location for C_R. In the former case, the generated eccentricity was defined as mass eccentricity (e_m) whereas in the latter case, the generated eccentricity was defined as stiffness eccentricity (e_s) (Tso and Myslimaj, 2003). In contrast, some of the researchers investigated strength eccentric structures by producing difference in the strength of LFRS to vary the position of C_V with relative to C_M (Kim *et al.*, 2012). In such cases, the generated eccentricity was called as strength eccentricity (e_v). The illustration corresponding to various eccentric systems are pictorially presented in Figures 1.2 and 1.3.

Interest in the research works on plan-asymmetric structures started growing rapidly from last three decades when Tso and Sadek (1985) evaluated the performance of mass eccentric one-story models by means of varying ductility demands in the mass eccentric structure. Results from the analytical research demonstrated the predominant effect of structural time-period on the ductility demands beyond elastic range. The research outcome demonstrated a total of 20% difference in the results obtained between the Clough and bilinear model. Later, Bozorgnia and Tso (1986) and Tso and Bozorgnia (1986) examined plan-asymmetric structures with e_s and e_v for the corresponding inelastic seismic response. Similarly, Sadek and Tso (1989) performed analytical investigations on the mono-eccentric systems to assess the inelastic response of the structures with strength eccentricities. From the analytical findings, the code-based stiffness eccentricity criteria was found to be useful in elastic response prediction. However, for the inelastic response, the criteria of strength eccentricity were found to be useful.

Pekau and Guimond (1990) used elasto-plastic force-deformation relationship to examine the effectiveness of accidental eccentricity (e_a) in order to account for the torsional induced vibrations under varying stiffness and strength of LFRS. The findings of the research demonstrated torsional enhancement in the seismic responses under the influence of e_s and e_v. Finally, the inadequacy of 5% provision for the accidental eccentricity was explained. Chandler and Duan (1991) proposed a revision in the design eccentricity in Mexico code-87 based on their analytical investigations on plan-irregular structures as listed in Table 2.1.

Table 2.1 Proposed revision in the design eccentricity (Chandler and Duan, 1991)

Eccentricity	Primary eccentricity	Secondary eccentricity
Proposed revision	$1.5e_s + b$	$0.5e_s - 0.1b$
Mexico code-87	$1.0e_s - 0.1b$	$1.0e_s - 0.05b$

Chandler and Hutchinson (1992) evaluated stiffness eccentric structures for potential torsional coupling by means of one-story models. Findings of the research demonstrated strong correlation between the time-period of the structure and torsional coupling. Moreover, the research findings demonstrated higher displacement demands at the FS when compared with SS. Chandler *et al.* (1995) further demonstrated and justified provisions of various design codes accounting for the torsion in asymmetric structures. The research was conducted on torsionally unbalanced (TU) and torsionally balanced (TB) models. TU models were established by introducing stiffness eccentricity in the models. The considered TU structures were designed such as to account for both moderate and low torsional stiffness. The designed models were named A1 and A2. Research findings pertaining to the varying stiffness eccentricities in the considered models concluded higher deformation demands at FS when compared with the deformation demands at the SS. For long period structures, the ductility demands were found to be higher whereas the SS of the models with shorter period was found to experience lesser ductility demands. However, ductility demands in the TU models exceeded by 2.5% when compared with TB models. Ferhi and Truman (1996a, 1996b) evaluated asymmetric structures with strength and stiffness eccentricities to determine their impact on the elastic and inelastic seismic response. The research findings demonstrated great dependence of seismic response on the e_s in the elastic range. However, the effect of e_v on seismic response was observed to be in the inelastic range. De-La-Colina (1999a, 1999b) investigated the influence of torsional coupling in TU structures under orthogonal seismic excitation components in the horizontal plane. The author investigated the following parameters: (a) time-period of the structure, (b) design eccentricity and (c) seismic force reduction factor. The research concluded that for TU-SS the ductility demands augmented with natural period, and the opposite was observed for TU-FS. Regarding the design eccentricity, the research demonstrated that the amplification in the stiffness eccentricity caused reduction in the ductility demands. Furthermore, the research concluded a reduction in the ductility demand at the FS of the asymmetric structure when the force reduction factor was increased. Overall, the research concluded that strength eccentricity has far higher influence on the seismic demands relative to the stiffness eccentricity. Ghersi and Rossi (2001) conducted elastic and inelastic seismic analyses under orthogonal earthquake components and presented a detailed comparison of the elastic and inelastic responses in stiffness eccentric structures. Research findings demonstrated that elastic analysis under unidirectional seismic excitation overestimated the engineering demands. Moreover, it has been explained that the use of bi-directional seismic excitation led to minor variation in the engineering demands. De Stefano and Pintucchi (2002) assessed plan-asymmetric structures with stiffness eccentricities and evaluated the interaction between the horizontal and axial forces in the inelastic range. The research demonstrated that the interaction between the horizontal and axial forces results in 20% reduction in the floor rotation demands. Moreover, it has been presented that the application of unidirectional earthquake component results in the lower deformation demands at both the FS and SS. Similar outcomes have also been identified by Myslimaj and Tso (2002). Research proposed by De Stefano and Pintucchi (2002) evaluated one-story structures considering the inelastic interaction between the orthogonal lateral force in the LFRS and the axial force. The research presented that previous studies on torsional influence generally overestimate the torsional response as such studies did not consider the interaction phenomena because it can lead to smaller floor rotations except for short-period structures. In Heredia-Zavoni and Machicao-Barrionuevo (2004), a linear asymmetric structure with bi-directional floor eccentricity was exposed to orthogonal components of seismic excitation in the horizontal plane in order to assess its influence on the seismic response. The research demonstrated strong dependence of the seismic response on the translational period of the structure and whether the asymmetric structure is TF or TS. Moreover, this research explained that seismic response is highly dependent on the soil conditions. Ladinovic (2008) presented base shear torque surface (BST) of strength and stiffness eccentric structure and described that parameters influencing the BST surface. These influencing parameters were mainly torsional capacity, lateral capacity and strength eccentricity. The BST surfaces were described as representative inelastic seismic response. Jarernprasert and Bazan (2008) and Jarernprasert *et al.* (2008) evaluated the one-story plan-asymmetric structures having stiffness eccentricities and compliance with the provisions of Mexico-Code (2004) and IBC (2009). The research demonstrated the influence of seismic excitations on the following characteristics: (a) normalized static eccentricity and structural elastic period, (b) design target ductility and (c) ratio of uncoupled torsional to transitional frequencies. It was concluded that the Mexico-Code (2004) overestimated the design forces at the FS of the structure whereas IBC (2009) leads to the overestimation of the design forces at both the SS and FS of the considered asymmetric structures. Aziminejad and Moghadam (2009) conducted nonlinear dynamic analysis on strength eccentric plan-asymmetric structures subjected to near-fault and far-fault earthquakes. It has been demonstrated that seismic response is highly sensitive to strength distribution in TS structures under both near-fault and far-fault earthquakes whereas in TF structures, the impact of distribution of strength on seismic response is negligible. The research concluded that the TS structures containing balanced position of C_V-C_R demonstrate better seismic performance under both near-fault and far-fault earthquakes compared with other structures considered in the research. Lucchini *et al.* (2010) verified the BST surface procedure using incremental dynamic analysis (IDA). One-story structures with bi-directional eccentricities were considered to establish the research findings.

2.3   Research on multi-story asymmetric structures

Despite the fact that one-story analytical models are the most simplified form of asymmetric structures, they are still attractive to researchers because of their potential to clarify effective and useful parameters for the seismic design criteria. However, in the past two decades, research on multi-story asymmetric structures has increasingly captured the interest of researchers for the following reasons:

- Establishment of various computational programs for more precise and effective analysis of 3D multi-story structures.
- Incapability of one-story models in estimating torsional seismic response of actual structures has been discovered by several authors (Stathopoulos and Anagnostopoulos, 2002, 2003).
- Simplified single-story models are not representative of multi-story structures where estimation of actual seismic response is more complicated.

In contrast with the previous studies on simplified one-story asymmetric structures, Stathopoulos and Anagnostopoulos (2005) attempted to investigate more refined multi-story structures and the corresponding torsional response in the nonlinear range. The research considered three- and five-story asymmetric structures designed in accordance with UBC (1997) and EC-8 (2005) for seismic response evaluation. The research findings concluded that the SS of the asymmetric structures experience lesser deformation demands compared with their regular counterpart whereas the FS of the structure may experience higher inelastic deformation demands. Marušić and Fajfar (2005) evaluated multi-story mass eccentric steel structures under orthogonal components of seismic excitation for elastic and inelastic seismic response. Similarly, Peruš and Fajfar (2005) demonstrated qualitative behavior of inelastic and elastic seismic responses. De Stefano *et al.* (2006) examined the effects of overstrength in element cross-sections by considering a code designed six-story structure exposed to unidirectional seismic excitation. The research demonstrated that in actual structures the ductility demand may entirely be different than the simplified one-story models. Particularly, it is demonstrated that FS of the asymmetric structures may lead to higher ductility demand in the upper floors compared to the SS of the structure.

Duan and Chandler (1997) further investigated TB and TU structures to establish an optimized method for seismic response of the structures under investigation. Ghersi and Rossi (2001) concluded that bi-directional ground motions result in little variation in the seismic response. Unidirectional ground motions overestimated the elastic seismic response. Shakib and Ghasemi (2007) considered plan-asymmetric structures with stiffness eccentricity and demonstrated the seismic behavior of asymmetric structures under the influence of near-field and far-field earthquakes. Numerous researchers have questioned the accuracy of torsional response in single-story asymmetric structures as such structures do not truly depict the behavior of multi-story asymmetric structures. De-La-Colina (2003) considered mass and stiffness eccentric multi-story structures under bi-directional ground motions and estimated optimal values of eccentricities. Penelis and Kappos (2005) incorporated the influence of translational and torsional modes in single and multiple degree-of-freedom structures and investigated the inelastic seismic response based on a new proposed method. De Stefano *et al.* (2006) considered one-story and multi-story structures and investigated the difference between their seismic responses. Aziminejad and Moghadam (2009) demonstrated the seismic performance of strength and stiffness eccentric multi-story structures with different distribution of strength. Anagnostopoulos *et al.* (2010) evaluated mass and stiffness eccentric multi-story structures and determined their inelastic torsional response. The seismic response for one of the considered models contradicted with the SS response of single-story structures. Bosco *et al.* (2004) investigated non-regularly asymmetric structures, characterizing different eccentricity levels in the structure and in plan-asymmetry over the entire elevation of the structure. Dimova and Alashki (2003) carried out analytical and numerical investigation on regular structures and concluded that even small eccentricities in symmetric structures lead to irregular behavior and excessive accidental torsional effects. Several other research studies on multi-story structures with horizontal and vertical eccentricities have been reported in the literature (Penelis and Kappos, 2005; Peruš and Fajfar, 2005; Fajfar *et al.*, 2005; Chopra and Goel, 2004; Chintanapakdee and Chopra, 2004; Chopra and Chintanapakdee, 2004; Das and Nau, 2003; Tremblay and Poncet, 2005; Karavasilis *et al.*, 2008; Athanassiadou, 2008; D'Ambrisi *et al.*, 2009; Van Thuat, 2013; Georgoussis, 2016; Lim *et al.*, 2018; Karayannis and Naoum, 2018; Eivani *et al.*, 2018; Fang and Leon, 2018).

2.4   Research based on development of analysis procedure for asymmetric structures

Various researchers proposed analysis procedures for asymmetric structures. One of the earlier studies in this regard was conducted by Duan and Chandler (1997) which demonstrated an optimized procedure pertaining to the seismic response of TU and TB systems. The authors proposed overstrength-factor and design eccentricity expressions with code defined expressions. The proposed expressions were found to be applicable to both multi-story and single-story TU systems. Similarly, in one of the earlier studies on proposed procedures for asymmetric structures, an alternative pushover procedure was proposed by Moghadam and Tso (2000). Dutta and Das (2002a, 2002b) proposed two hysteresis models accounting for the stiffness and strength deterioration of LFRS under cyclic loading and implemented the proposed models in single-story plan-asymmetric structures exposed to orthogonal components of earthquakes. The research concluded that both the FS and SS of the plan-asymmetric structures showed variation when strength deterioration was considered. Tso and Myslimaj (2003) presented a novel analytical method corresponding to the stiffness and strength distribution in the asymmetric structure. Based on the dynamic analysis, balanced location of C_V-C_R was determined in the considered models. The proposed method demonstrated satisfactory results in terms of determining balanced location of C_V-C_R and lower torsional response. Research conducted by Fujii et al. (2004) proposed a simple method for plan-asymmetric systems with stiffness eccentricities considering non-linear analysis. The research demonstrated that TS structures may likely experience higher vibrations in the first mode compared with the TF-structures. Moreover, comparing the seismic response of SDOF and MDOF for TF and TS structures, it was found that SDOF models were applicable to TS structures only. The proposed procedure efficiently determined the seismic response of TS structures. Aziminejad and Moghadam (2009) utilized the research method of Tso and Myslimaj (2003) to investigate the performance of asymmetric structures with stiffness, strength and mass eccentricities. The research concluded that the best location of C_V and C_R is dependent upon the required performance level. Trombetti and Conte (2005) proposed a simple procedure for the estimation of maximum rotational response. The proposed method was also verified in another study for plan-asymmetric structures Trombetti et al. (2008).

2.5   Experimental work on asymmetric structures

Despite the fact that numerous researchers have conducted analytical research on planar and vertically asymmetric structures, experimental research on these asymmetric structures is, however, very limited, and there is a very limited record available which describes the experimental evaluation of multi-story asymmetric structures. One of the earliest experimental studies on vertically asymmetric structure is reported in the literature (Shahrooz and Moehle, 1990a, 1990b) where the influence setback on static and dynamic response is investigated and the adequacy of seismic design provision has been evaluated.

Investigation of asymmetric structures under ground motion accelerations has always been a complex problem. Lately, some experimental works have been done to capture the seismic behavior of such structures (McCrum and Broderick, 2013; Bousias et al., 2007). Lee and Ko (2004) evaluated three 17-story scaled RC structures with three types of irregularities in the structure. The asymmetric structures were designed according to the provisions of Korean code and the research was conducted by means of shake table testing. The experimental findings described low drift demands in lower floors. However, these lower drift demands were accompanied by no significant influence on the reduction of overturning moment, base shear and overturning deformations. Similarly, bi-directional pseudo-dynamic tests (Negro et al., 2004) on actual scale, three-story asymmetric structures were conducted to assess the seismic response of asymmetric structures (Jeong, 2005). The research findings based on the experimental testing demonstrated the torsional influence under intense seismic shaking. Furthermore, recent investigation by De-La-Colina *et al.* (2007) presented experimental investigation on eight one-story asymmetric steel structures under progressive seismic excitations to investigate the elastic and inelastic response of the structure. The research presented torsional amplification-factors for appropriate design of the asymmetric structures. Other research studies on the experimental investigation of asymmetric structures have been reported in the literature (Moehle and Alarcon, 1986; Lee *et al.*, 2011; Lee and Hwang, 2015; Zhang *et al.*, 2017; Yenidogan *et al.*, 2018; Richard *et al.*, 2016).

2.6   Seismic design provisions for asymmetric structures

The current seismic design practice for asymmetric structures is by means computing the design eccentricities to cater for the influence of torsion. The expressions for the design eccentricities (e_d^1, e_d^2) have been developed such that to include both inherent eccentricity and accidental eccentricity (percentage multiple of the plan dimension). The reason for considering the accidental eccentricity in the design process is to account for various unpredictable factors that may have not been addressed in the design process. For instance, some of these unpredictable factors are the effects of seismic rotational component and spatial distribution of load. It should be noted that the accidental eccentricity is multiplied by a factor A_x to account for the adverse cases of asymmetry. These design eccentricities contribute towards the amplified response for FS of the structure as presented in Equation 2.1 whereas for the SS of the structure, the design eccentricity leads to a reduced factor as explained in Equation 2.2:

$$e_d^1 = \alpha e_s + A_X(e_a) \tag{2.1}$$

$$e_d^2 = \delta e_s - A_X(e_a) \tag{2.2}$$

Where e_d^1 refers to the primary design eccentricity and e_d^2 is the secondary design eccentricity. In Equations 2.1 and 2.2, the term e_a refers to the accidental design eccentricity and equals to the multiple of a coefficient and plan dimension, L. Therefore, the accidental eccentricity can be expressed as:

$$e_a = \beta L \tag{2.3}$$

The terms α, δ, β and A_X are the amplification coefficient for the FS of the structure, amplification coefficient for the SS of the structure, percentage coefficient for accidental eccentricity and torsional irregularity coefficient respectively. Different codes specify different values of these coefficients.

A summary of provisions for design eccentricities adopted by different seismic design codes including ASCE/SEI (2016), EC-8 (2005), NBCC (2005) and IBC (2009) is presented in Table 2.2.

In Table 2.2, it can be seen that EC-8 (2005) has its own procedure for the approximate analysis accounting for the effects of torsion in asymmetric structures. The term refers to the radius of gyration and depends upon the geometric properties of the structure as has been defined in Equation 2.4:

$$\gamma = \sqrt{\frac{\left(L^2 + B^2\right)}{12}} \tag{2.4}$$

Table 2.2 Comparison of design eccentricities adopted by various international design codes

Seismic design codes	Primary design eccentricity (e_d^1)	Secondary design eccentricity (e_d^2)
IBC (2009)	$1.0e_s + A_x(0.05L)$	$1.0e_s - A_x(0.05L)$
ASCE/SEI (2016)	$1.0e_s + A_x(0.05L)$	$1.0e_s - A_x(0.05L)$
EC $-$ 8 (2005) $\begin{cases} \\ \\ \\ \end{cases}$	$\left(1.0 + \dfrac{e_2}{e_s}\right)e_s + 0.05L$ $e_2 = 0.1(L+B)\sqrt{\dfrac{10e_2}{L}} \leq 0.1(L+B)$ $e_2 = \dfrac{1}{2e_0}\left[\gamma^2 - e_0{}^2 - K_\theta^2 + \sqrt{\left(\gamma^2 - e_0{}^2 - K_\theta^2\right)^2 + 4e_0{}^2 K_\theta^2}\right]$	$1.0e_s - 0.05L$
NBCC (2005)	$1.5e_s + 0.10L$ $0.5e_s + 0.10L$	$0.5e_s - 0.10L$ $1.5e_s - 0.10L$

Table 2.3 Design code provisions for asymmetric structures

Irregularity	IBC (2009)	EC-8 (2005)	ASCE/SEI (2016)	NBCC (2005)
Stiffness (KI)	$K_i < 0.7K_{i+1}$ or $K_i < 0.8(K_{i+1} + K_{i+2} + K_{i+3})$	$K_i < 0.7K_{i+1}$ or $K_i < 0.8(K_{i+1} + K_{i+2} + K_{i+3})$	$K_i < 0.7K_{i+1}$ or $K_i < 0.8(K_{i+1} + K_{i+2} + K_{i+3})$	$K_i < 0.7K_{i+1}$ or $K_i < 0.8(K_{i+1} + K_{i+2} + K_{i+3})$
Mass (MI)	$M_i < 1.5\,M_a$	Abrupt reduction is restricted	$M_i < 1.5\,M_a$	$M_i < 1.5\,M_a$
Soft story (SI)	$K_i < 0.7K_{i+1}$ or $K_i < 0.8(K_{i+1} + K_{i+2} + K_{i+3})$	–	$K_i < 0.7K_{i+1}$ or $K_i < 0.8(K_{i+1} + K_{i+2} + K_{i+3})$	$K_i < K_{i+1}$
Weak story (WI)	$K_i < K_{i+1}$	–	$K_i < 0.6K_{i+1}$ or $K_i < 0.7(K_{i+1} + K_{i+2} + K_{i+3})$	–
Re-entrant corners (CI)	–	$C_i \leq 5\%$	$C_i \leq 15\%$	–

Similarly, corresponding to the design eccentricities, different seismic design codes prescribe different limits of irregularities. A summary of various characteristics of irregularities is presented in Table 2.3 (Varadharajan *et al.*, 2012).

Where KI, MI, SI, WI and CI refer to stiffness irregularity, mass irregularity, soft story irregularity, weak story irregularity and re-entrant corner irregularity respectively. Moreover, in Chapter 8, to represent strength irregularity, the term VI will be used. The first term in these abbreviations defines the structural parameter and the second term stands for irregularity. For instance, in strength irregularity (VI), V stands for shear strength and I stands for the irregularity. It can be seen from the limits described in Table 2.3, that the majority of the seismic design codes prescribe similar guidelines for the irregularities which are based on the magnitude of irregularity, ignoring the aspect of location-specific influence of irregularity on the seismic response, which is unrealistic. Moreover, the past earthquake records show poor seismic performance of such irregular structures during earthquakes. As can be seen in Table 2.3, the structural irregularities can be of various types. To understand the classification of these irregularities, they have been presented in Figures 2.1 and 2.4.

From the literature study, it is evident that accurate estimation of seismic response is complicated mainly because of the following reasons:

- Seismic response is influenced by the characteristics of seismic excitations.
- Seismic response is affected by the mode excited in the structure.
- Seismic response is influenced by the type of structural inherent eccentricity.
- Various seismic demands have different influence under the same eccentricity and same seismic excitation.

2.7   Influence of seismic excitation characteristics

Post-earthquake studies have demonstrated structural damage even for engineered structures (Osteraas and Krawinkler, 1989; Ger et al., 1993; Minzheng and Yingjie, 2008). Such damage to the structure is mainly contributed by the lack of understanding towards the behavior of asymmetric structures, damage causing nature of the earthquakes (Ni et al., 2013; Wen et al., 2014; Raghunandan and Liel, 2013; Ruiz-García et al., 2014) and uncertainty in the seismic excitation (Takewaki, 2005; Li et al., 2004). Therefore, precise prediction of engineering demand is what the current seismic design practice lacks. Due to the described factors, precise prediction of the engineering demand is considered a very complicated process because of the multicomponent nature of the seismic excitation and because of the fact that seismic excitation can hit the structure along any direction. This in turn leads to the uncertainty in the seismic response associated with the varying orientations of seismic excitations. Numerous researchers have demonstrated the influence of varying orientations of ground motion on elastic and inelastic seismic responses. For elastic seismic response, various analytical formulae are developed by Athanatopoulou (2005) to investigate the critical seismic response of three correlated seismic components. The research concluded that critical direction of a ground motion changes with the response quantity of interest and characteristics of seismic excitation. These conclusions have also been presented in Kalkan and Kwong (2013) and Alam et al. (2017) where the influence of orientation of seismic excitation on various response quantities have been illustrated based on a linear 3D structure. Kostinakis et al. (2008) examined the critical orientation of seismic excitation and the corresponding peak response quantity on the basis of the formulae (Athanatopoulou, 2005) for special classes of buildings subjected to isotropic bi-directional ground motion (Kalkan and Kwong, 2013). The seismic motion is recorded in the planar and vertical direction. The recorded components of earthquake are generally correlated. However, Penzien and Watabe (1974) identified the existence of uncorrelated seismic components which could be used to determine the critical orientation of an earthquake. The determination of critical seismic response can be obtained using the response-spectrum method. Therefore, Smeby and Der Kiureghian (1985) estimated the critical-orientation of the seismic excitation with the same spectral shape. López and Torres (1997) calculated the peak response quantities corresponding to critical orientations based on the response spectrum procedure for seismic excitations with three components with identical and different spectral shape. Menun and Kiureghian (1998) calculated the critical orientation and the corresponding response quantities based on CQC3 rule. Lopez *et al.* (2000, 2001) presented a study and concluded that the peak response value at critical angle for a single response quantity could be 20% more compared with the response obtained when the ground motions are considered along reference axes of the structure. The response corresponding to the critical orientation for the most unfavorable combinations of various engineering demand parameters was determined by Menun and Kiureghian (2000a, 2000b).

Previous research concludes that orientation of the seismic excitation significantly affects the seismic response, and the critical angle depends upon the characteristics of ground motion and response quantity of interest Kalkan and Kwong (2013) and Alam *et al.* (2017). Most of the previous studies revolve around the estimation of maximum response through critical angle or through combination rules. However, to what extent the earthquake directionality can influence the performance of an asymmetric structure is still a question. Also, the uncertainty of the seismic responses at the SS and FS independently hasn't yet been studied in detail.

2.8   Research based on damage/failure assessment of asymmetric structures

A building structure may collapse or suffer severe damage under the action of seismic forces due to sudden change in the stiffness/strength and mass in the plan or in the vertical direction of the structure as such kind of changes are the primary triggers for damage in the structure. In regards to this issue, it is evident from the literature review that experimental investigation on damage assessment of asymmetric structures is very limited (Wang, 2018). The majority of the research on damage response of asymmetric structures is based on the post-earthquakes damage assessment of case-study structures. The next subsections discuss a few such examples reported in the literature.

2.8.1   Damage in plan-asymmetric structures

Numerous major and minor earthquakes in the past have demonstrated damage to the asymmetric structures. The planar irregularities can be classified based on irregular geometric shape, re-entrant corners, diaphragm discontinuity, mass, strength and stiffness irregularities. These classifications are illustrated in Figure 2.1.

These planar irregularities may lead to poor performance of the plan-asymmetric structures during the event of an earthquake and cause various kinds of damage depending upon the type of planar irregularity. For instance, shear failure in the column under global system

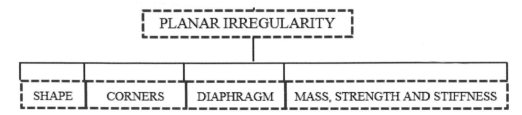

Figure 2.1 Classification of plan-asymmetric structures

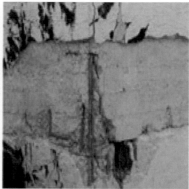

Figure 2.2 Shear failure in the column under global system torsion (Yön et al., 2017 and http://db.world-housing.net/)

torsion can be seen in Figure 2.2 where a plan-asymmetric structure formed additional shear effects under torsional vibrations caused by stiffness eccentricity.

Figure 2.3 shows the Ministry of Culture building which was damaged due to torsion during the Haiti earthquake in 2010. The presence of asymmetric stiffer region in the structure resulted in torsion which led to damage of LFRS away from the C_R of the structure. Due to collapse of flexible elements, the entire floor experienced a downward pull, which led to the total collapse of the building. Because of the poor seismic design provisions, structural damage is likely to happen.

A structural engineer requires good understanding regarding the behavior of plan-asymmetric structures in order to avoid potential failure of the structure during major or minor seismic events. Since most of the past research works mainly focused on the simplified asymmetric systems, there is a need for a comprehensive evaluation on the effects of interaction of various irregularities on the seismic response parameters to formulate improved design philosophy for these structures.

(a)

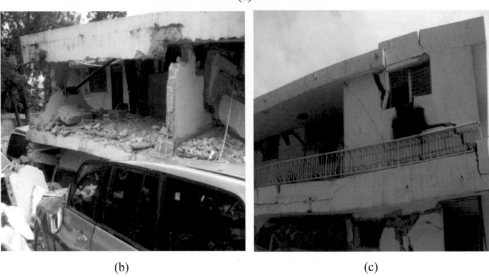

(b)   (c)

Figure 2.3 Damage in stiffness eccentric structure under Haiti earthquake 2010 (courtesy of Mid America Earthquake Engineering Center, University of Illinois at Urbana Champaign): (a) overall damaged structure, (b) damage at the flexible edge and (c) damage at the stiff edge

2.8.2   *Damage in vertical-asymmetric structures*

Vertical-asymmetry refers to the asymmetric distribution of stiffness/strength and/or mass along the height of the structure and can be classified into various types, such as vertical mass-asymmetry, stiffness-asymmetry, strength-asymmetry and setbacks. An illustration of these asymmetric classifications is presented in Figure 2.4.

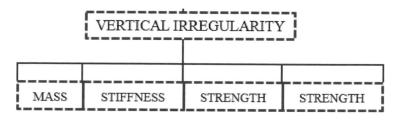

Figure 2.4 Classification of vertical-asymmetric structures

(a)

(b)

Figure 2.5 Failures in asymmetric structures: (a) soft story mechanism (Yön et al., 2017) and (b) weak story collapse mechanism (Yön et al., 2017)

One of the vertical asymmetry most commonly present in the real-life structures is the presence of soft story, which is the one of the most influencing irregularities for the collapse of vertically asymmetric structures (Adalier and Aydingun, 1998; Huang and Skokan, 2002; Sezen et al., 2003). Likewise, in Figure 2.5, various asymmetric structures with a soft/weak story are presented. The presented structures experienced a higher amount of damage under extreme torsional vibrations. Under these vibrations, the shear stress concentrations increased in the columns of the building located on the FS of the structure and caused failure of the entire floor. The illustrations of the failures caused by the soft/weak story mechanism are also reported in Figure 2.5.

The presence of soft-story mechanism leads to higher deformation demands during a seismic event eventually causing the soft-story columns to dissipate the whole seismic energy. Much structural damage can be associated with higher deformations in conjunction with the poorly addressed torsional influence. The soft story has been one of the major reasons of damage throughout the world during earthquakes as evident from seismic reports. Therefore, it is prescribed to avoid sudden change of stiffness/strength and mass in the vertical direction of the structure (Fig. 2.6).

Figure 2.6 Typical earthquake damage in asymmetric structures (http://db.world-housing.net/)

2.8.3 Damage assessment based on shake table testing

The majority of the previous research deals with the damage assessment of case-identified asymmetric structures. Therefore, the conclusions developed from those studies are restricted to the circumstances corresponding to the damages in the case-study asymmetric structures and are not likely to be applicable for various asymmetric structures. This eventually demands the damage/collapse-based study of asymmetric structures using shake table testing for a wide range of representative asymmetric scenarios. However, in terms of shake table testing of asymmetric structures for damage evaluation it is evident from the literature survey that very limited research has been done in the past for both global response estimation and local damage evaluation of asymmetric structures. For instance, Kim *et al.* (2012) conducted experimental investigation on an asymmetric structure having stiffness and strength eccentricities and assessed the structural capacity under torsional induced vibrations.

Similarly, Benavent-Climent *et al.* (2014) carried out experimental investigations on scaled RC structures and evaluated the influence of torsion under uniaxial seismic excitations. Since shake table testing on asymmetric structures for damage investigation is very limited. Numerous research studies are based on the mechanism of damage at component level. For instance, Wu *et al.* (2005) conducted experimental research on one-story structures to investigate the collapse behavior of asymmetric structures. Wu *et al.* (2009) further extended experimental research to investigate flexure shear failure and axial failure modes. The research findings were then used for the development of simplified methods to assess the damage response of asymmetric structures. Similarly, Elwood and Moehle (2008b) conducted experimental investigation on one-story RC structures and verified the axial failure and shear failure modes of the columns using nonlinear analyses (Elwood and Moehle, 2008a). The experiment was designed such as to redistribute the exterior columns load path once the interior column failed and lost its full capacity. Ghannoum and Moehle (2012) carried out a shaking table experiment on a three-story RC-frame structure to investigate the local failures in the columns. The research findings were then used to investigate the global failure as a result of local failure in the structure. Similarly, Li *et al.* (2016) conducted experimental investigations to assess the damage process in an RC frame structure under input seismic waves. The results from these studies are specific to these circumstances and don't cover a wide range of characteristics of asymmetric structures. In this regard, a detailed investigation on the damage behavior of asymmetric structures for both local responses and global behavior is required. However, measuring the accurate damage inside the concrete under dynamic testing for the prediction of cracks and damage assessment is a complicated task as the traditional strain gauges are only helpful in determining the damage state on the surface of concrete with numerous other limitations, which are explained in the next section. One technique for the accurate assessment of damage inside the crack is the embedment of fiber Bragg grating (FBG) strain sensors. However, previous literature on the study of FBG-based damage investigation in asymmetric structures under a dynamic test is nearly negligible. The majority of the research concerning the use of FBG sensors for damage evaluation is limited to other civil infrastructure facilities. This research is, therefore, dedicated to the use of FBG strain sensors in asymmetric structures for dynamic testing to assess its efficiency and suitability for dynamic testing of asymmetric structures.

2.8.4   Applications of FBG strain sensors for damage assessment

As discussed earlier, the damage investigation studies in the past used traditional methods of measurements such as placing a traditional strain gauge onto the surface of concrete to evaluate the strain response under dynamic actions. Such methods are incapable of demonstrating the actual damage response inside the structural components and thus are not considered sophisticated enough for more precise measurements. In the recent past, more advanced damage monitoring procedures have been established which are more sophisticated and more accurate. One such measurement method is through the installation of FBG sensors (Rao, 1999; Majumder *et al.*, 2008; Ye *et al.*, 2014) inside the concrete members. In this regard, several researchers conducted experimental investigations to demonstrate the effectiveness of FBG sensors. For instance, Brönnimann *et al.* (1998) investigated a cable-stayed bridge partly under construction and demonstrated the long-term applications of FBG sensors. Similarly, Kister *et al.* (2007) carried out research on strain monitoring in pile foundations by means of FBG sensors. Further studies on the damage response of an RC bridge for inelastic deformations were conducted by Kerrouche *et al.* (2008b, 2008a). Research investigations corresponding to the applications of FBG sensors in different civil infrastructure facilities have been contributed by various researchers (Rodrigues *et al.*, 2012; Barbosa *et al.*, 2008; Kerrouche *et al.*, 2009; Chan *et al.*, 2006; Chung *et al.*, 2008). Similarly, numerous researchers have developed novel FBG sensors for monitoring purposes (Lin *et al.*, 2005b; Costa and Figueiras, 2012; Zhou *et al.*, 2011; Lin *et al.*, 2005a, 2006). The earlier-described literature review clearly indicates that the use of FBG sensors is very widespread (Kim *et al.*, 2013; Arsenault *et al.*, 2013; Bang *et al.*, 2012; Ye *et al.*, 2013; Li *et al.*, 2008, 2009; Biswas *et al.*, 2010; Chan *et al.*, 2000; Naruse *et al.*, 2000; Liu *et al.*, 2016; Schulz *et al.*, 2001; Mita and Yokoi, 2001; Ren *et al.*, 2006b, 2006a; Sun *et al.*, 2015).

In terms of applications of FBG sensors in the damage investigation of building structures, very few studies are on record (Ni *et al.*, 2009; Li *et al.*, 2012). This demands the investigation corresponding to the applications of FBG sensors in the damage evaluation of building structures, and therefore, this research has considered FBG sensors for the damage evaluation of experimental models described in Chapter 3 of this book.

2.8.5   Research gap in terms of damage assessment in asymmetric structures

Previous literature related to FBG-based damage investigation of RC structures does not fully guide us about the damage response in building structures, and the main reason behind this is that the majority of FBG-based health monitoring studies are limited to bridges, dams, tunnels and other infrastructure facilities. Its applications are not as widespread to buildings as its applications to other infrastructure facilities (Ni *et al.*, 2009). Moreover, previous research on FBG sensors provides an insight into the structural performance with apparently no physical damage to the structure. Estimating the damage response of a physically damaged structure is yet another complicated process and highly depends upon the type sensing technique and location of the sensors. It should be noted that despite the extensive use of FBG sensors in various infrastructure facilities, there is nearly no previous research on FBG sensor-based damage response in asymmetric building structures. Therefore, there is a need to fill this research gap by detailed damage investigation in building structures using FBG sensors where a structure actually transforms from an elastic state to a highly inelastic state under the formation of cracks followed by progressive seismic excitations.

It should also be noted that damage investigation of RC buildings by traditional methods does not truly depict the building's damage response. The main reason behind this is that the usual practice for damage response monitoring in a shaking table test of a structure involves the use of traditional strain-based damage measurement methods where resistance strain gauges are used to measure the strain of structures. These strain gauges are pasted onto the surface of concrete through adhesives. These instant adhesives for fixed strain gauges can react with model material, and eventually they can influence the measurement accuracy. Besides, they are also vulnerable to having reduced insulation resistance under intense shaking as they are likely to get removed or displaced, followed by the damage in the structure and eventually the accuracy of the measurement gets compromised. Some other key points that further enhance the novelty of this research are as follows:

- As explained previously, the research on FBG-based damage response asymmetric buildings undergoing inelastic deformations is nearly nonexistent. Therefore, FBG strain sensors were used in this study as these sensors have the ability of periodical variation in the index of refraction of the optical fiber core (Kersey et al., 1997; Li et al., 2004). Because of its numerous advantages over other technologies and its appropriate features, which include its embedding abilities, high sensitivity, electro-magnetic interference immunity and flexibility, its application in successfully investigating the damage process is evaluated.
- The existing data stock on experimental damage response of asymmetric structures is not significant; therefore, data collected in this research can be used for future evaluation of damage response, verification and development of computational models for prediction of damage behavior in asymmetric RC structures.

2.8.6   Research gap in terms of global behavior of asymmetric structures

For seismic performance investigation of asymmetric structures, some researchers evaluated one-story models whereas others considered multi-story structures. Single-story models were largely used by previous researchers with the majority of recent researchers preferring multi-story building models in comparison to the multi-story building models with few exceptions (Anagnostopoulos et al., 2010; Stathopoulos and Anagnostopoulos, 2010). The expressions proposed in seismic design codes (regarding torsional irregularity) is based on single-story SDOF systems and elastic analysis (Aziminejad and Moghadam, 2009). Thus, the code provisions are not valid for multi-story building models. Therefore, there is a necessity to revise these provisions. Moreover, the modified expressions need to be generalized to have wider applicability. One of the major problems with previous studies on the global response investigations is that they were not capable of demonstrating the precise structural response along with the seismic design guidelines (Myslimaj and Tso, 2002; Tso and Myslimaj, 2003; Aziminejad and Moghadam, 2009).

Therefore, broadly considering the issues with previous studies, the following research gaps can be identified:

- The influence of irregularity is dependent on the variation of seismic response parameters and the location-specific influence of irregularity. Except for a few numerical investigations (Dutta and Das, 2002a, 2002b), the research on the influence of location-specific asymmetry is nearly none.
- The majority of the observations are based on the conclusions derived from the evaluation of single-story asymmetric structures.
- Previous research studies are only based on the numerical investigations, and the experimental investigations of asymmetric structures are very limited.
- Asymmetric structures have not been investigated for both local damage and global behavior simultaneously.
- The majority of the previous numerical investigations are based on nonlinear static procedures. However, these procedures do not truly capture the influence on seismic response under varying characteristics of seismic excitations. Therefore, capturing the true and effective influence of asymmetry, seismic response under varying characteristics of ground motion components over the entire duration of an earthquake time history is important to consider for global response evaluations.
- Previous studies are limited to deformation or ductility demands. The influence of torsion on various seismic demand parameters such as rotation at the FS and SS of an asymmetric structure is still not known very well.
- Behavior of local and global response of the asymmetric structure at the SS and FS has neither been explored nor been correlated under the interaction of irregularities.
- Seismic design guidelines pertaining to the interaction of irregularities are not available even in the seismic design codes.

2.9   Summary

From a detailed investigation of the previous literature, it has been found that both qualitative and quantitative demonstrations on the influence of one-story and multi-story asymmetric structures have been evaluated by numerous researchers (Stathopoulos and Anagnostopoulos, 2002, 2003, 2005; Anagnostopoulos *et al.*, 2010; Kyrkos and Anagnostopoulos, 2011; Anagnostopoulos *et al.*, 2015b, 2015a). However, the presented research investigations are not yet helpful in improving the current seismic design practice due to the reasons explained in previous sections.

Moreover, eccentricity variation in an asymmetric structure under a seismic event provides critical information in terms of torsional contribution towards seismic response. This is because of the fact that seismic response is not only influenced by the characteristics of seismic excitation but the variation in the eccentricity as well. Additionally, there is a need to investigate the influence of varying eccentricity on each individual seismic demand and at both the FS and SS separately. This will help practicing engineers to develop an understanding for the response transition from the FS to the SS or from the SS to the FS at both component level and global level. Therefore, keeping in view the research gap, it seems imperative that extensive experimental research is required on a wide range of asymmetric structures to establish conclusions on the seismic response based under a complete dynamic process.

Chapter 3

Experimental strategy and seismic loading program

3.1 Introduction

This chapter provides details of the experimental strategy: geometric configuration of the experimental models, material details, instrumentation and the characteristics of seismic loading used for the experimental work. However, before demonstrating the details of the experimental setup, it is imperative to demonstrate how the classification of the experimental structures was made in terms of the asymmetric structure being TS or TF. Figure 3.1 demonstrates the static eccentricity as the difference of C_S and C_M. The original design of all the experimental models was planned such as to possess either inherent stiffness, strength or both the irregularities altogether. Moreover, these models were designed such that to possess these irregularities either in floor plan or elevation or both altogether. The geometry of the building was defined on the basis of the parameter Ω. This parameter is the key parameter which influences the behavior of an asymmetric structure. The parameter Ω depends upon the uncoupled frequency of the structure and is described as the ratio of the uncoupled rotational frequency to the uncoupled translational frequency. To consider this parameter, a simplified building model is assumed as illustrated in Figure 3.1.

The n-story structure is approximated as n-structures with 3-DOF. The equivalent stiffness is approximated by the stories below the ith level, while the equivalent mass is approximated by the stories above the ith level. Hence, the global stiffness in the plan around the vertical axis is considered to estimate the approximate locations of the center of stiffness.

Each system with 3-DOF has some location of C_S ($\beta x, CR, \beta_{y, CR}$) at each story level. Defining the location of C_S in multi-story structures is a very complex problem (Hejal and Chopra, 1989a; Fujii et al., 2004; Bousias et al., 2007). The global C_S of the structure can be located using the weighted average of the C_S of the floor below the ith level. Using the story height h_i and the floor stiffness, the respective weights can be evaluated as follows (Cimellaro et al., 2014):

$$\beta x,\ CR = \left(\sum_{i=1}^{n} \alpha_{xi,\ CR} \cdot k_{li} \cdot \left(h_i \right)^3 \right) \Big/ \left(\sum_{i=1}^{n} k_{li} \cdot \left(h_i \right)^3 \right) \tag{3.1}$$

$$\beta y,\ CR = \left(\sum_{i=1}^{n} \beta_{yi,\ CR} \cdot k_{ti} \cdot \left(h_i \right)^3 \right) \Big/ \left(\sum_{i=1}^{n} k_{ti} \cdot \left(h_i \right)^3 \right) \tag{3.2}$$

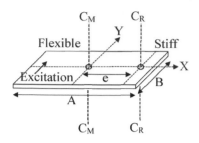 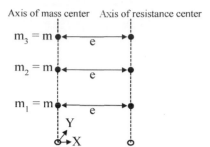

Figure 3.1 Schematic representation of asymmetry

The approximate global frequencies of the nth 3-DOF system can be expressed as:

$$\omega_{x,\,G} = \left(K_x \,/M\right)^{0.5} \tag{3.3}$$

$$\omega_{y,\,G} = \left(K_y \,/M\right)^{0.5} \tag{3.4}$$

$$\omega_{\varnothing,\,G} = \left(K_\varnothing \,/\,(M.\ e^2 + J'_{\varnothing,G})\right)^{0.5} \tag{3.5}$$

In Equation 3.5, $J'_{\varnothing,G}$ refers to global-polar moment of inertia and can be expressed as:

$$J_{\varnothing,G} = \left(\sum_{i=1}^{n} J'_{\varnothing,i} + m_i \left(\left(\alpha_{xmi} - \alpha_{xG}\right)^2 + \left(\alpha_{ymi} - \alpha_{yG}\right)^2\right)\right) \tag{3.6}$$

In Equation 3.6, $J'_{\varnothing,i}$ refers to polar moment of inertia of the respective story at mass center and α_{xG} and α_{yG} are the global center of gravity coordinates of the nth-DOF system. Similarly, the term Ω described earlier can be expressed as follows:

$$\Omega = \left(\omega_{\varnothing,\,G} \,/\omega_{x,\,G}\right)^{0.5} \text{ and } \Omega = \left(\omega_{\varnothing,\,G} \,/\omega_{y,\,G}\right)^{0.5} \tag{3.7}$$

For the considered experimental models with A/B ratio of 2 and 1, and an eccentricity between C_M and C_S which equals 0.45A-1A, the damage and global torsional behavior of the asymmetric structure is quantified. In this study, the type of experimental structure as either TF or TS is established based on the global quantities described in Equations 3.1–3.7.

Experimental models in this book are broadly classified into two categories: (1) regularly irregular (RI) structure and (2) irregularly irregular (IRI) structure. RI structures refer to the state of irregularity where all floors have uniform eccentricity whereas IRI structures refer to the state of irregularity where all floors have different irregularity randomly distributed. IRI state has been included in this book to observe the influence of interaction of irregularities on both local damage response as well as global response.

3.2   Experimental models

Because of the varying material characteristics of the concrete, it is not possible to investigate a wide range of parameters even with multiple concrete structures of the same type. There are several reasons behind this limitation. For instance, there are various construction and manufacturing issues which will not let the concrete structures possess uniform characteristics in all models and hence the dynamic characteristics of the constructed models may entirely differ from each other despite the fact they possess similar geometry and material. Besides, it is very challenging to maintain uniform structural characteristics of the concrete model for each variation of the eccentricity once the concrete structure is made to experience seismic excitation. The reason behind this is that even for very small amplitude seismic loading, concrete structures are expected to form micro-cracks which eventually influence the dynamic characteristics of the structure. One possible solution to overcome this issue is to simulate the damage characteristics using steel models for a wide range of vertical and planar irregularities. Therefore, to simulate both the damage characteristics and global behavior for various parameters corresponding to the asymmetric behavior of the structures, steel models along with the RC model were designed and constructed. The RC frame model is named C-1 whereas the steel moment resistant frame models are named S-1, S-2, S-3 and S-4, respectively.

3.2.1 RC model: C-1

The RC frame-shear wall structure named the C-1 model was constructed to evaluate and predict the physical damage characteristics inside the structural components under seismic torsional vibrations induced to the eccentricity between C_M and C_R. The presented RC model in Figure 3.2 is intended to provide damage characteristics as well as global structural

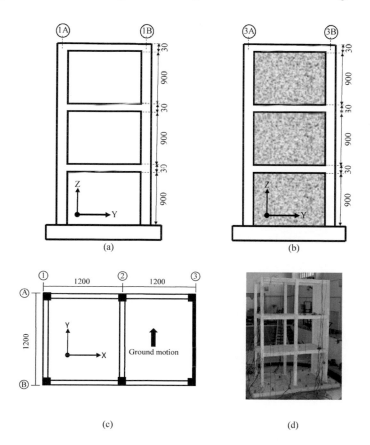

Figure 3.2 C-1 model; dimensions in mm: (a) elevation; Y-frame, flexible side; (b) elevation; Y-frame, stiff side; (c) floor plan and (d) 3D model

behavior in both elastic and inelastic states under the progressive increase of seismic excitations. This structure is considered an RI model in the elastic state and an IRI model in inelastic state due to non-uniform yielding in the structure. The IRI state in the inelastic state is considered for the fact the non-uniform yielding in the structure tends to shift the C_M of the structure. This shift in the C_M occurs randomly at different floors and consequently leads to non-uniform irregularity along the height of the structure.

3.2.2   Steel model: S-1

These models consist of two square-shape three-story steel structures and are shown in Figure 3.3 along with their regular counterpart. These are TS structures with one of them being mono-directional plan-asymmetric while the other one being bi-directional plan-asymmetric. The C_M in these structures is designed to be located at the C_G of the structure while C_R was displaced from the C_G to form a normalized stiffness eccentricity of 0.45 ($e_s/L = 0.45$) at all floor levels.

Considering each floor of the S-1 models as a combination of the independent floor system, the uncoupled torsional frequency ratio for each independent floor system is greater than unity and thus, the S-1 models are considered TS models. The described state of the S-1 model where all floor levels have the same normalized stiffness eccentricity of 0.45 is termed as reference state of S-1 models. Since in the reference state the S-1 models have the regular floor eccentricity along the height of the structure, the models have the characteristics of regular plan-asymmetry or in other words, the described state of the models is considered as RI state.

3.2.3   Steel model: S-2

This model consists of one square-shape three-story steel structure and is represented in Figure 3.4 along with its regular counterpart. It is a TS structure containing bi-directional

(a)   (b)   (c)

Figure 3.3 S-1 models: (a) counter-symmetric model, (b) bi-directional asymmetric S-1 model and (c) mono-symmetric S-1 model

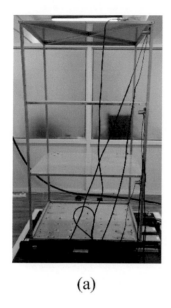

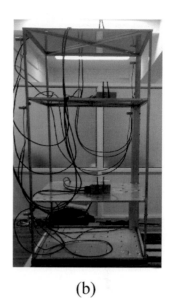

(a)   (b)

Figure 3.4 S-2 model: (a) counter-symmetric model and (b) bi-directional asymmetric S-2 model

eccentricities at all floor levels. Moreover, this model also consists of vertical asymmetry in the form of the weak/soft story at second- and third-floor levels as the model is designed such that to comply with the limits of weak/soft floor defined in Table 2.3. This was achieved by reducing the columns' stiffness and increasing the floor height at the second-floor level to 1.5 times of adjacent lower and upper floors. For the third-floor level, the columns' stiffness was further reduced. However, floor height was not increased at the third-floor level. Therefore, the S-2 model has non-uniform stiffness and strength eccentricities at all floor levels. Considering the S-2 model a combination of three independent floor systems, the uncoupled torsional frequency ratio for each independent floor system is greater than unity and thus, the S-2 model is considered a TS model. The described state of the S-2 model where all floor levels have the non-uniform strength and stiffness eccentricity is termed as reference state of S-2 model. Since in the reference state the S-2 model has the irregular floor eccentricities in vertical direction, therefore, the model in its reference state is described to possess IRI state of irregularity.

3.2.4   Steel model: S-3

This model is an L-shape three-story steel structure having bi-directional plan-asymmetry and re-entrant corners at all floor levels as illustrated in Figure 3.5 along with the regular counterpart. The S-3 model also has the same normalized stiffness eccentricity of 0.45 ($e_s/L =$ 0.45) at all floor levels. The re-entrant corners of the S-3 model were designed to violate the limit defined in Table 2.3. Considering the S-3 model a combination of three independent floors, the uncoupled torsional frequency ratio for each independent floor system turns out to be greater than unity and therefore, the S-3 model is considered a TS model. The described

(a)   (b)

Figure 3.5 S-3 model: (a) counter-symmetric model with re-entrant corners and (b) bi-directional asymmetric structure with re-entrant corners

state of the structure where all floor levels have the same normalized stiffness eccentricity of 0.45 is defined as reference state of S-3 model. Since in the reference state the S-3 model has the regular floor eccentricity in vertical direction, therefore, the model is an RI model in its reference state.

3.2.5   Steel model: S-4

This model is a circular-shape asymmetric model with both planar and vertical irregularities as shown in Figure 3.6. The structure was designed and constructed such as to contain bi-directional floor eccentricities. The S-4 model has non-uniform stiffness and strength eccentricities at all floor levels. Considering each floor of the S-4 model as a combination of the independent floor system, the uncoupled torsional frequency ratio for each independent floor system is less than unity and thus, the S-4 model is considered as a TF model. The described state of the S-4 model where all floor levels have the non-uniform strength and stiffness eccentricity is termed as reference state of S-4 model. Since in the reference state the S-4 model has the irregular eccentricities along the height of the structure along with different planar eccentricities at each floor level, the model is considered an IRI model. Moreover, the intermediate floor of the S-4 model was constructed to be 1.5 times higher than the adjacent lower and top floors along with reduced column section at the SS of the intermediate floor. This was done to produce the effects of the weak/soft story at the intermediate floor level as the intermediate floor stiffness is approximately 0.6 times the adjacent lower and top floor stiffness.

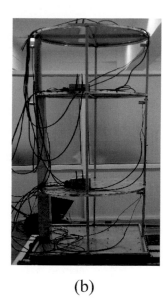

(a)   (b)

Figure 3.6 S-4 model: (a) counter-symmetric model and (b) bi-directional asymmetric circular model

3.3   Eccentricity variation in the experimental models

The planar and vertical regularity of a structure is an important control index of structural seismic design, which eventually has an impact on the seismic response of asymmetric structure. In this regard, the shake table experiments were conducted considering the earlier-defined reference state of each model. However, to include various asymmetric cases, the C_M of each model at floor level simultaneously or separately varied to increase the eccentricity between C_M and C_S or C_M and C_V or all of them together.

3.3.1   Variation of floor eccentricity in RC model

For the RC model, it should be noted that it is not possible to produce a controlled floor eccentricity as the structure transformed from an elastic state into the inelastic state following non-uniform yielding of the structural components. Therefore, it has been assumed for the C-1 model that the change in the eccentricity is associated with the varying damage condition of the model as the C_M of the RC model changes during the complete dynamic process.

3.3.2   Variation of eccentricity in steel models

The eccentricities in each of the S-1 through S-4 models were varied by shifting the C_M. Therefore, to simulate the damage characteristics and global behavior of these models for various asymmetric conditions, 24 different asymmetries were generated in each model and each asymmetric model was tested under four different seismic excitations. However, these asymmetries have broadly classified into nine cases and have been reported in Table 3.1.

Table 3.1 Asymmetric cases of steel models

Case	Characteristics of eccentricity	Sub-case	Eccentricity variation on the floor(s) of interest
1	Reference state	–	$e_s/L = 0.50$
2	1. Planar and vertical irregularities 2. Mass and stiffness eccentricity variation at the first-floor level 3. Stiffness eccentricity at second- and third-floor level: $e_s/L = 0.50$ 4. Median response	Case 2A Case 2B Case 2C Case 2D Case 2E	$M_2, M_3 < 1.2 M_1$ and $e_s/L = 0.55$ $M_2, M_3 < 1.4 M_1$ and $e_s/L = 0.60$ $M_2, M_3 < 1.6 M_1$ and $e_s/L = 0.65$ $M_2, M_3 < 1.8 M_1$ and $e_s/L = 0.70$ $M_2, M_3 < 2.0 M_1$ and $e_s/L = 0.75$
3	1. Planar and vertical irregularities 2. Mass and stiffness eccentricity variation at the second-floor level 3. Stiffness eccentricity at first- and third-floor level: $e_s/L = 0.50$ 4. Median response	Case 3A Case 3B Case 3C Case 3D Case 3E	$M_1, M_3 < 1.2 M_2$ and $e_s/L = 0.55$ $M_1, M_3 < 1.4 M_2$ and $e_s/L = 0.60$ $M_1, M_3 < 1.6 M_2$ and $e_s/L = 0.65$ $M_1, M_3 < 1.8 M_2$ and $e_s/L = 0.70$ $M_1, M_3 < 2.0 M_2$ and $e_s/L = 0.75$
4	1. Planar and vertical irregularities 2. Mass and stiffness eccentricity variation at the third-floor level 3. Stiffness eccentricity at first- and second-floor level: $e_s/L = 0.50$ 4. Median response	Case 4A Case 4B Case 4C Case 4D Case 4E	$M_1, M_2 < 1.2 M_3$ and $e_s/L = 0.55$ $M_1, M_2 < 1.4 M_3$ and $e_s/L = 0.60$ $M_1, M_2 < 1.6 M_3$ and $e_s/L = 0.65$ $M_1, M_2 < 1.8 M_3$ and $e_s/L = 0.70$ $M_1, M_2 < 2.0 M_3$ and $e_s/L = 0.75$
5	1. Planar irregularities 2. Mass and stiffness eccentricity variation at first, second and third-floor level 4. Median response	Case 5A Case 5B Case 5C Case 5D Case 5E	$e_s/L = 0.55$ $e_s/L = 0.60$ $e_s/L = 0.65$ $e_s/L = 0.70$ $e_s/L = 0.75$
6	Center of mass and center of stiffness converged at one point but dislocated from the geometric center of the structure	–	Normalized eccentricity = 0
7	1. Third floor's mass 3 times higher than the adjacent lower floors 2. Constant stiffness eccentricity at first- and second-floor level: $e_s/L = 0.50$	–	$M_1, M_2 < 3 M_3$
8	1. Second floor's mass 3 times higher than the adjacent upper and lower floor 2. Constant stiffness eccentricity at first- and third-floor level: $e_s/L = 0.50$	–	$M_1, M_3 < 3 M_2$
9	1. First floor's mass 3 times higher than the adjacent upper floors 2. Constant stiffness eccentricity at second- and third-floor level: $e_s/L = 0.50$	–	$M_2, M_3 < 3 M_1$

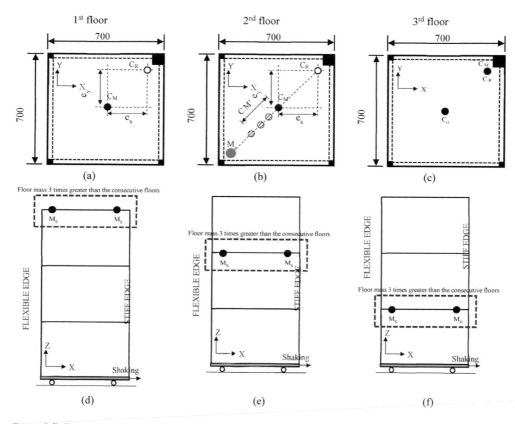

Figure 3.7 Demonstration of a few asymmetric cases in steel models: (a) RI-bi-eccentric, (b) IRI-bi-eccentric, (c) RI with dislocated C_M and C_R, (d) top floor's mass 3 times greater than the adjacent floor's mass, (e) intermediate floor's mass 3 times greater than the adjacent floor's mass and (f) first floor's mass 3 times greater than the adjacent floor's mass

The representative cases of eccentricities can also be observed in Figure 3.7.

3.4   Design of the experimental models

Model design is one of the important issues that were required to be addressed especially for RC model in order to obtain successful results. Therefore, the design of the RC model was not only based on determining the similarity constants but various other factors such as the type of production model, material of the model, the test conditions and a model to determine the physical similarity constants. The model test theory is based on the similarity principle and the dimension analysis to determine the similarity criteria that must be followed in the model design. The process of structural model testing objectively reflects the interrelationship between the relevant physical quantities involved in the work. Because of the similarity between the prototype and the model, it necessarily reflects the relationship between the similarity of the prototype structure and the model. The similarity constant will thus determine the relationship between the prototype structure and model, that is, the model design needs to follow these principles. The model test not only requires the elastic phase of the stress analysis of the data, but also requires a correct reflection of the nonlinear structure of the prototype performance requirements which can then be reflected to the prototype structure, i.e. the ultimate deformation capacity and the ultimate bearing capacity, which is more important for the structural seismic test.

The similarity here refers to the correlation between prototype and the model, which is broader than the usual geometric similarity concept. The so-called physical phenomena are similar, meaning that the ratio between the physical quantities of the first and second processes is constant at the corresponding time in the entire system in which the physical process is performed, in addition to the geometric similarity.

The stress and strain of the designed model according to the scale ratio were kept consistent with the prototype which eventually means that the displacement of the model is consistent with the prototype. According to the scale reduction, the testing instrument required a much higher accuracy. This condition is generally more difficult to satisfy because it is difficult to maintain the damping coefficient constant because of the change in the structural properties under progressive loading. This condition can be ignored if the prototype structure damping is assumed to be small. The model was designed at ¼ scale; the required vibration frequency of the table was 4 times the frequency of the earthquake. The amplitude of the vibration table was the seismic amplitude. This is because the configuration model according to the scale ratio of the frequency itself was increased four times and the displacement was reduced and the test of the vibration table also made the corresponding changes to meet the model test results similar to the prototype.

The structures were designed for peak ground acceleration of 0.28g with a 5% damped design spectrum using Chinese structural design code GB50011-2001. The Chinese seismic design code adopts three levels of seismic design PGAs. The first PGA level refers to 63% probability of exceedance in 50 years (minor event), the second level PGA refers to 10% probability of exceedance in 50 years (moderate event), and the third level PGA refers to the 2%–3% probability of exceedance in 50 years (rare event). The quantified definitions of these three levels of hazards are reported in Table 3.2.

The GB50011-2001 code requires the structural system to behave elastically under a minor event, and not be significantly damaged under a moderate and/or rare event. For the Shanghai area as the interested site here, the corresponding peak acceleration for the rare event was considered as 0.28g in this research. The reinforcement design of the plan-asymmetric structure was carried out in accordance with the Chinese structural design code

Table 3.2 Hazard quantification

Seismic event	Hazard quantification	
	Probability of exceedance in 50 years (%)	Return period (years)
Minor	63	50
Moderate	10	475
Rare	2–3	1640–2475

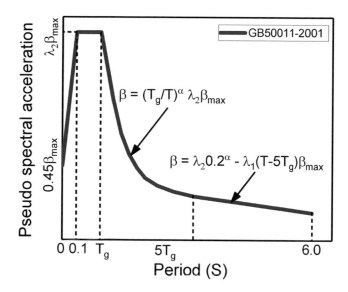

Figure 3.8 Seismic design spectrum

(GB50011-2001). The seismic design spectrum is illustrated in Figure 3.8 and was established using Equations 3.8–3.11.

$$
S_a(T) = \begin{cases}
\beta_{max}\left(0.45 + \dfrac{\lambda_2 - 0.45}{0.1}T\right), & 0 \leq T \leq 0.1 \\[2mm]
\lambda_2\beta_{max}, & 0.1s \leq T \leq T_g \\[2mm]
\lambda_2\beta_{max}\left(\dfrac{T_g}{T}\right)^\alpha, & T_g \leq T \leq 5T_g \\[2mm]
\beta_{max}\left[\lambda_2 0.2^\alpha - \lambda_1\left(T - 5T_g\right)\right], & 5T_g \leq T \leq 6
\end{cases}
\tag{3.8}
$$

where λ_1, λ_2 and α are the coefficients determined from the following equations:

$$
\lambda_1 = 0.02 + \frac{0.05 - \zeta}{8}, \quad \lambda_1 \geq 0
\tag{3.9}
$$

$$
\lambda_2 = 1 + \frac{0.05 - \zeta}{0.06 + 1.7\zeta}, \quad \lambda_2 \geq 0
\tag{3.10}
$$

$$
\alpha = 0.9 + \frac{0.05 - \zeta}{0.5 + 5\zeta}
\tag{3.11}
$$

In the preceding equations, β_{max} depends upon the seismic design level of the structure, which in turn depends upon the peak ground acceleration, and seismic fortification intensity of the site of interest. Based on the guidelines of GB50011-2001, the design spectrum was established for seismic fortification intensity of VIII (0.2g–0.3g), maximum influencing coefficient of horizontal seismic intensity 0.28g (2% in 50 years) with a design character-istics period of ground motion being 0.45s and a soil profile type II (140–250 m/s). The spectrum was chosen to conform to the code defined earthquake for the Shanghai region.

3.4.1   Design of RC model

Keeping in view the stringent requirements of reinforced concrete structure and the actual situation available for the testing facility, similarity relationships were established. The structural properties of high-rise buildings under seismic actions are usually carried out on a shaking table using the same scale model as the prototype model. That is, the physical process contains the following physical quantities: structure size, the level of structural dis-placement, stress, strain, the elastic modulus of the structural material, the average density of structural materials, the weight of the structure, the vibration frequency of the structure and structural damping ratio along with the displacement amplitude of seismic excitation and the maximum frequency of movement. Using the dimensional analysis method, the similarity constants of the system were defined as the dimension matrix.

The experiment on the C-1 model was performed at the State Key Laboratory of Coastal and Offshore Engineering, Dalian University of Technology, China. Keeping in view the limited capacity of the available shake table facility, the best possible scale of the structure was decided to be 1:4. The shake table consists of 3m × 4m dimensions having the ability to excite the model in uni-direction with a payload capacity of 10 tons. The occupants' masses were not incorporated in this study due to the limited payload capacity of the shake table. Table 3.3 refers to the similitude requirements implemented in this book where N demon-strates the ratio of prototype properties and model properties. Stress similarity factor N_σ and

Table 3.3 Similarity relationships for RC model

Properties	Parameter	Similarity equations	Dimensions	Scale factor
Geometric properties	Length (l)	N_l	L	0.25
	Lateral displacement (y)	$N_y = N_l$	L	0.25
Material properties	Strain (ε)	$N_\varepsilon = 1$		1
	Stress (σ)	$N_\sigma = N_E N_\varepsilon$	FL^{-2}	1
	Modulus of elasticity (E)	$N_E = N_\sigma$	FL^{-2}	1
	Density (ρ)	$N_\rho = N_\sigma / N_l$	FT^{2}L^{-4}	2.165
	Poison ratio (μ)	$N_\mu = 1$	–	1
Dynamic properties	Mass (m)	$N_m = N_\rho N_l^3$	FT^{2}L^{-1}	0.033
	Period (T)	$N_T = \sqrt{N_m/N_K}$	T	0.367
	Frequency (f)	$N_f = 1/N_T$	T^{-1}	2.719
	Velocity (v)	$N_v = N_y/N_t$	LT^{-1}	0.67
	Acceleration (a)	$N_a = N_y/N_t^2$	LT^{-2}	1.848
	Acceleration of gravity (g)	$N_g = 1$	LT^{-2}	1
Load	Surface load (Q)	$Nq = N\sigma$	FL^{-2}	1

Table 3.4 Similarity relationships for steel models

Properties	Parameter	Similarity equations	Dimensions	Scale factor
Geometric properties	Length (l)	N_l	L	0.166
	Lateral displacement (y)	$N_y = N_l$	L	0.166
Material properties	Strain (ε)	$N_\varepsilon = 1$		1
	Stress (σ)	$N_\sigma = N_E N_\varepsilon$	FL^{-2}	1
	Modulus of elasticity (E)	$N_E = N_\sigma$	FL^{-2}	1
	Poison ratio (μ)	$N_\mu = 1$	–	1
Dynamic properties	Acceleration (a)	$N_a = N_y / N_t^2$	LT^{-2}	2.77
	Acceleration of gravity (g)	$N_g = 1$	LT^{-2}	1

length similarity factor N_l are considered the key controlling parameters for defining the similitude relationships.

3.4.2 Design of steel models

The experiments on the S-1 through S-4 models were performed at Structural Vibration Control Laboratory, Qingdao University of Technology, China. Keeping in view the limited capacity of the available shake table facility, the best possible scale of these structures was decided to be 1:6. Since the steel models were constructed to provide the seismic response details over wide characteristics of different asymmetric models, the similarity of these models became irrelevant as the asymmetric conditions were induced by introducing heavy mass at various locations. However, in the reference state of the structure, these models were designed using the design spectrum criteria illustrated in Figure 3.8. The similarity criteria for the design of the steel models are reported in Table 3.4.

3.5 Material and geometric details of the experimental models

For conventional test conditions, factors that were carefully looked after were: materials and premise of the construction model, the eccentric design framework of a two-way cross one-way shear wall model in case of the C-1 model and eccentric condition of steel models, as illustrated in Figures 3.2–3.6. Integrally cast stratified model construction methods were used for the C-1 model and the model was left for 28 days to achieve maximum strength of concrete. For the fabrication of steel models, the beam and column elements were welded together to form the structure.

3.5.1 Material and geometric details of RC model

The prototype model was based on integrally cast-in stratified model construction methods. The prototype was left for 28 days after construction in order to achieve the maximum strength of concrete during this period. Structural frame components were constructed having longitudinal reinforcement bars of diameter 6.5mm and steel wire stirrups. The base of

Table 3.5 Mechanical properties of reinforcement

Dia. of reinforcement (mm)	Elastic modulus (GPa)	Tensile strength (MPa)	Yield strength (MPa)
6.5	200	276	350
2	200	276	350

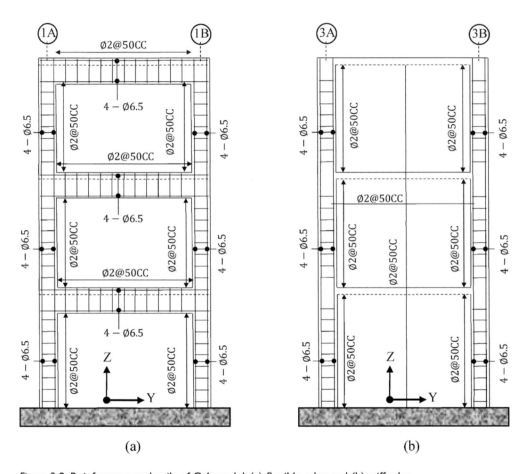

(a)   (b)

Figure 3.9 Reinforcement details of C-1 model: (a) flexible edge and (b) stiff edge

the structure was constructed with C30 concrete, and the frame components were constructed using M5 pouring mortar. The properties of the reinforcement are described in Table 3.5.

The model design, geometric configuration and reinforcement description is illustrated in Figure 3.9. The prototype model was constructed with columns of 80mm × 80mm and with beams of size 40mm × 100mm. The slab and shear wall was considered to be 30mm thick. The overall dimension of the structure is 2.4m in the longitudinal direction and 1.2m in the transverse direction and the height of the structure is 1.2m at all floor levels. The structure was designed in accordance with the design spectrum criteria presented in Figure 3.8.

The longitudinal reinforcing bars in beams and columns were of diameter 6.5mm and steel wires were used as shear stirrups. The concrete material was used as C20 concrete using M5 pouring mortar. All columns throughout the structure's height and in all frames were similar (with the same reinforcement, height and cross-sectional area), and so were the beams in all structural frames as shown in Figure 3.2. The beam sections were rectangular in shape with a size of 40mm × 100mm, and the columns were of square shape with a size of 80mm × 80mm.

3.5.2 Material and geometric details of steel models

The geometric and material characteristics of steel models are listed in Tables 3.6–3.8. It should be noted that in these models beam-column joints were welded together to form relatively rigid joints. The column and beam sections were designed to have a section size of 15mm × 15mm except at the SS, where the column section varied in size in each steel model.

Table 3.6 Geometric and material characteristics of models S-1 and S-3

Component	Thickness (mm)	Elastic modulus (MPa)	Section size (mm)
Beam	2	2×10^5	15 × 15
FS column	2	2×10^5	15 × 15
SS column	2	2×10^5	80 × 80
Slab	1.5	2×10^5	–

Table 3.7 Geometric and material characteristics of model S-2

Component	Thickness (mm)	Elastic modulus (MPa)	Section size (mm)
Beam	2	2×10^5	15 × 15
FS column	2	2×10^5	15 × 15
	2	2×10^5	80 × 80 (first floor)
SS column	2	2×10^5	60 × 60 (second floor)
	2	2×10^5	40 × 40 (third floor)
Slab	1.5	2×10^5	–

Table 3.8 Geometric and material characteristics of model S-4

Component	Thickness (mm)	Elastic modulus (MPa)	Section size (mm)
Beam	1	2×10^5	15
FS column	1	2×10^5	15 × 15
SS column	1	2×10^5	60 × 60 (first floor)
	1	2×10^5	40 × 40 (second floor)
	1	2×10^5	20 × 20 (third floor)
Slab	0.75	2×10^5	–

It should be noted that for the described irregular models S-1 through S-4, counter regular models were constructed. These regular models were constructed to evaluate the enhanced response in asymmetric structures. Figures 3.2–3.6 represent the constructed models where balanced models are representative regular parts of asymmetric models of each shape. A total of eight steel models were established for experimental investigation which also includes the counter regular models. The structural models were constructed considering a scale ratio of 1/6 having an aspect ratio of 1 and a size of 700mm × 700mm along X and Y directions. The height of these structures is 500mm at all floor levels and in all structural models except for S-2 and S-4 where the second-floor height is 800mm. Each asymmetric was constructed with a counter symmetric model, and the asymmetric models were constructed such that to contain both planar and vertical irregularities eventually leading to formation of the FS and SS in the structures as shown in Figures 3.2–3.6. Two of the models were constructed with re-entrant corners (Figures 3.4a and b), where Figure 3.4a corresponds to a balanced model having C_M and C_S converged at one point. The last two models are circular in shape with one of it being asymmetric and the other one is the corresponding symmetric model (Figures 3.6a and b).

The beam and column elements in the steel models consist of hollow box rectangular sections having dimensions 15mm × 15mm unless otherwise specified in Tables 3.6–3.8. The floor consists of 1mm thick steel plate in all structural models except S-4, where the steel plate thickness consists of 0.75mm. The change in the column size in the S-2 and S-4 models along the height of the structure was made to induce default vertical stiffness/strength eccentricity in the structure. It should be noted that these structures were constructed so as to range from highly TS to highly TF.

3.6 Instrumentation of experimental models

The experimental models were carefully equipped with instruments at strategic locations. The scheme of installed instruments in the models is represented as follows.

3.6.1 Instruments used in RC model

The survival rate of the sensor is critical in damage detection of the structure. In addition to being careful in the deployment process, several protection measures were taken. Since the construction process of the reinforced concrete structure is a typical operation, the fiber grating sensors were carefully handled during the pouring of concrete. To ensure proper contact of the FBG sensor with reinforcement bars and to avoid the strain error caused by the strain transfer, the rebars were first polished with the grinding machine and then were polished with sandpaper so that the sensor could smoothly stick with the rebars as shown in Figure 3.10.

In order to prevent the polished crumbs from oil and other pollutants, cotton balls were dipped in alcohol to clean the polished places. The fiber grating sensor along the longitudinal arrangement of rebars had the issues related to sensor buffer and moisture, therefore, at first, glue was fixed around the fiber grating, then it was wrapped with epoxy resin and external winding gauze. The procedure corresponding to the protection process and layout of the sensor is shown in Figures 3.11–3.13.

The transmission fiber was vulnerable to various construction impacts like concrete pouring, vibration, the impact from the inner mould etc. Therefore, the length of the transmission

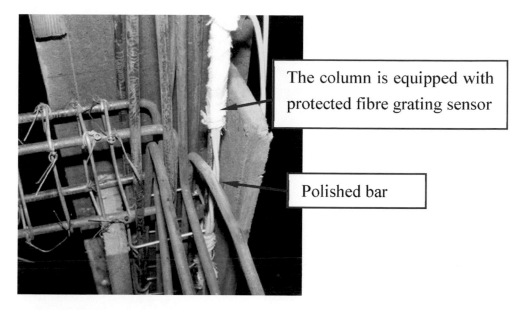

The column is equipped with protected fibre grating sensor

Polished bar

Figure 3.10 Polished reinforced steel bars

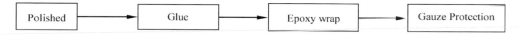

Polished → Glue → Epoxy wrap → Gauze Protection

Figure 3.11 Protection process of FBG sensor

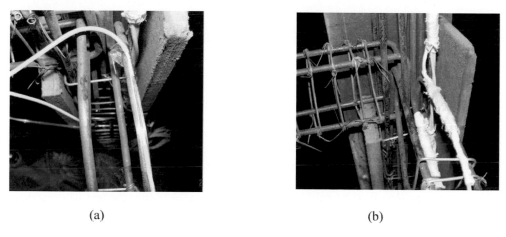

(a)   (b)

Figure 3.12 Protection measures for FBG sensors: (a) glued FBG sensor and (b) FBG sensor wrapped with gauze

Figure 3.13 Transmission fiber

fiber was kept as minimal as possible. A hole was left in the formwork near the fiber grating sensor so that the transmission fiber could pass through it easily. The laboratory environment was favorable for this to be done. Careful handling was done with the transmission fibers as they were not required to be excessively bent; otherwise, it could have led to a weak optical signal or even no signal.

The monitoring data from a total of 17 sensors were available after the test. The sensors were equipped with a total of 20 FBG strain sensors; all of them survived but since some of the sensors were single head and were not able to be connected to the channel, one of them was abandoned and the other one was a single head and was destroyed during the installation. For the single head sensor, the acquisition process was destroyed. Table 3.9

Table 3.9 FBG sensor layout plan in RC model

Sensor no.	Location of the sensor
1	The sensor is attached to the first rebar of the first story column located at the intersection of grid-1 and grid-A. The location of the sensor is the bottom of the column.
2	The sensor is attached to the second rebar of the first story column located at the intersection of grid-1 and grid-A. The location of the sensors is the bottom of the column.
3	The sensor is attached to the second rebar of the first story beam located at the intersection of grid-1 and grid-A. The location of the sensor is the first rebar of the transverse-direction beam at the first-floor level.
4	The sensor is attached to the second rebar of the second story column located at the intersection of grid-1 and grid-A. The location of the sensor is the second rebar of the transverse-direction beam at the second-floor level.
5	The sensor is attached to the first rebar of the third story column located at the intersection of grid-1 and grid-A. The location of the sensor is the first rebar of the transverse-direction beam at the third-floor level.
6	The sensor is attached to the first rebar of the first story beam located at the intersection of grid-1 and grid-B. The location of the sensor is the first rebar of the transverse-direction beam.
7	The sensor is attached to the second rebar of the first story beam located at the intersection of grid-1 and grid-B. The location of the sensor is the first rebar of the transverse-direction beam.
8	The sensor is attached to the first rebar of the second story beam located at the intersection of grid-1 and grid-B. The location of the sensor is the first rebar of the transverse-direction beam.
9	The sensor is attached to the first rebar of the second story column located at the intersection of grid-2 and grid-B. The location of the sensor is the first rebar of the transverse-direction beam.
10	The sensor is attached to the first rebar of the longitudinal-direction beam where grid-1 intersects with grid-A at the first-floor level.
11	The sensor is attached to the first rebar of the transverse-direction beam where grid-1 intersects with grid-A at the first-floor level.
12	The sensor is attached to the first rebar of the longitudinal-direction beam where grid-1 intersects with grid-B at the first-floor level.
13	The sensor is attached to the first rebar of the transverse-direction beam where grid-1 intersects with grid-B at the first-floor level.
14	The sensor is attached to the first rebar of the second story column located at the intersection of grid-1 and grid-A. The location of the sensor is the first rebar of the second floor's column.
15	The sensor is attached to the first rebar of the second story beam located at the intersection of grid-1 and grid-A. The location of the sensor is the first rebar of the transverse-direction beam.
16	The sensor is attached to the shear wall longitudinal reinforcement at the second-floor level at the junction of shear wall and floor.
17	The sensor is attached to the shear wall longitudinal reinforcement at the third-floor level at the junction of shear wall and floor.

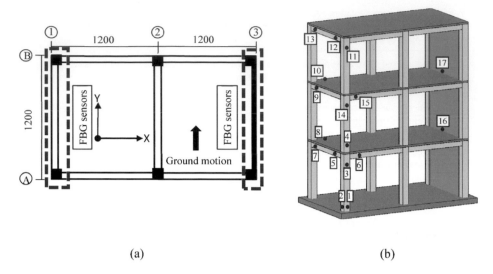

(a)   (b)

Figure 3.14 Layout plan of FBG sensors in RC model: (a) grids plan; the marked edges correspond to the regions of FBG sensor installation and (b) location of FBG sensors illustrated in 3D structure; the listed numbers correspond to the serial number of each sensor

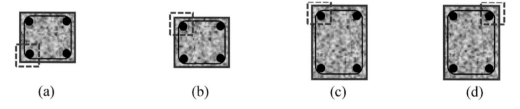

(a)   (b)   (c)   (d)

Figure 3.15 Location of FBG sensors in RC components: (a) first rebar of the column, (b) second rebar of the column, (c) first rebar of the beam and (d) second rebar of the beam

accompanied by Figures 3.14 and 3.15 illustrates the layout of FBG sensors in the test structure.

3.6.2   Instruments used in steel models

The deployed instruments to monitor the angular drift at the top roof level, accelerations at all floor levels and strain at all floor levels under dynamic shaking are shown in Figure 3.16. For a better comparison of the seismic responses, the instrumentations were arranged at both the FS and SS of the structure so as to monitor the response at both the edges of the structure.

The FBG strain sensors can be classified into two types: bare FBG sensor and tube-packaged FBG sensor. The tube-packaged FBG sensors were used in the concrete model test. However, for these extensive experiments, six bare FBG sensors were used in all models, among which three were used on each of the FS and SS of the structures. The arrangement of these FBG sensors is summarized in Table 3.10 (where $\varepsilon(t)$ refers to strain response in the time domain). The subscripts and superscripts refer to the corresponding floor number and edge of the structure, respectively.

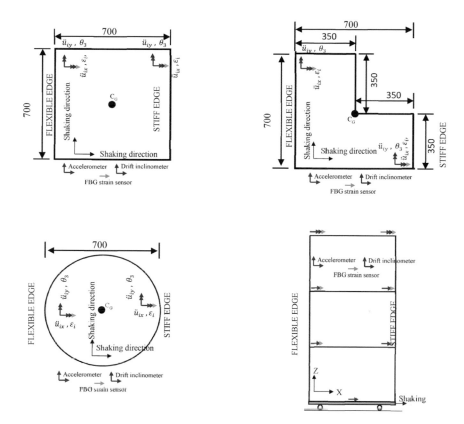

Figure 3.16 Layout of instruments in steel models

Table 3.10 Bare FBG sensors and their arrangement in steel models

Floor no.	Flexible side (FS)	Stiff side (SS)
Third floor	$\varepsilon(t)_3^F$	$\varepsilon(t)_3^S$
Second floor	$\varepsilon(t)_2^F$	$\varepsilon(t)_2^S$
First floor	$\varepsilon(t)_1^F$	$\varepsilon(t)_1^S$

Similarly, 12 unidirectional piezoelectric accelerometers were used at the FS and SS of the TU structures. Of these, six piezoelectric accelerometers correspond to each of the SS and FS of the structures. These accelerometers were arranged so as to acquire acceleration in the orthogonal plan at all floor levels. Three bi-directional accelerometers were placed at the geometric center of the first-, second- and third-floor levels of these structures. One accelerometer fixed in the shake table was used to obtain the acceleration data at the C_G of the base level of the structure. The summary of the arranged accelerometers has been reported in Table 3.11 (where $\ddot{u}(t)$ refers to the acceleration response in the time domain). The first term in the subscript defines the corresponding floor number whereas the alphabet refers to the response direction. The superscript refers to the corresponding edge or location of placement of the accelerometers.

Table 3.11 Piezoelectric accelerometers and their arrangement in steel models

Floor no.	Flexible side (FS)		C_G	Stiff side (SS)	
	X-direction	Y-direction		X-direction	Y-direction
S-1 (bi-directional asymmetric)					
Third floor	$\ddot{v}(t)^F_{3,x}$	$\ddot{v}(t)^F_{3,y}$	$\ddot{v}(t)^{C,G}_{3,x}$ and $\ddot{v}(t)^{G,C}_{3,y}$	$\ddot{v}(t)^S_{3,x}$	$\ddot{v}(t)^S_{3,y}$
Second floor	$\ddot{v}(t)^F_{2,x}$	$\ddot{v}(t)^F_{2,y}$	$\ddot{v}(t)^{G,C}_{2,x}$ and $\ddot{v}(t)^{G,C}_{2,y}$	$\ddot{v}(t)^S_{2,x}$	$\ddot{v}(t)^S_{2,y}$
First floor	$\ddot{v}(t)^F_{1,x}$	$\ddot{v}(t)^F_{1,y}$	$\ddot{v}(t)^{G,C}_{1,x}$ and $\ddot{v}(t)^{G,C}_{1,y}$	$\ddot{v}(t)^S_{1,x}$	$\ddot{v}(t)^S_{1,y}$
Base	–	–	Base sensor	–	–
S-2 (mono-symmetric)					
Third floor	$\ddot{v}(t)^F_{3,x}$	$\ddot{v}(t)^F_{3,y}$	$\ddot{v}(t)^{C,G}_{3,x}$ and $\ddot{v}(t)^{G,C}_{3,y}$	$\ddot{v}(t)(t)^S_{3,x}$	$\ddot{v}(t)^S_{3,y}$
Second floor	$\ddot{v}(t)^F_{2,x}$	$\ddot{v}(t)^F_{2,y}$	$\ddot{v}(t)^{G,C}_{2,x}$ and $\ddot{v}(t)^{G,C}_{2,y}$	$\ddot{v}(t)(t)^S_{2,x}$	$\ddot{v}(t)^S_{2,y}$
First floor	$\ddot{v}(t)^F_{1,x}$	$\ddot{v}(t)^F_{1,y}$	$\ddot{v}(t)^{G,C}_{1,x}$ and $\ddot{v}(t)^{G,C}_{1,y}$	$\ddot{v}(t)^S_{1,x}$	$\ddot{v}(t)^S_{1,y}$
Base	–	–	Base sensor	–	–
S-2					
Third floor	$\ddot{v}(t)^F_{3,x}$	$\ddot{v}(t)^F_{3,y}$	$\ddot{v}(t)^{C,G}_{3,x}$ and $\ddot{v}(t)^{G,C}_{3,y}$	$\ddot{v}(t)^S_{3,x}$	$\ddot{v}(t)^S_{3,y}$
Second floor	$\ddot{v}(t)^F_{2,x}$	$\ddot{v}(t)^F_{2,y}$	$\ddot{v}(t)^{G,C}_{2,x}$ and $\ddot{v}(t)^{G,C}_{2,y}$	$\ddot{v}(t)^S_{2,x}$	$\ddot{v}(t)^S_{2,y}$
First floor	$\ddot{v}(t)^F_{1,x}$	$\ddot{v}(t)^F_{1,y}$	$\ddot{v}(t)^{G,C}_{1,x}$ and $\ddot{v}(t)^{G,C}_{1,y}$	$\ddot{v}(t)^S_{1,x}$	$\ddot{v}(t)^S_{1,y}$
Base	–	–	Base sensor	–	–
S-3					
Third floor	$\ddot{v}(t)^F_{3,x}$	$\ddot{v}(t)^F_{3,y}$	$\ddot{v}(t)^{C,G}_{3,x}$ and $\ddot{v}(t)^{G,C}_{3,y}$	$\ddot{v}(t)^S_{3,x}$	$\ddot{v}(t)^S_{3,y}$
Second floor	$\ddot{v}(t)^F_{2,x}$	$\ddot{v}(t)^F_{2,y}$	$\ddot{v}(t)^{G,C}_{2,x}$ and $\ddot{v}(t)^{G,C}_{2,y}$	$\ddot{v}(t)^S_{2,x}$	$\ddot{v}(t)^S_{2,y}$
First floor	$\ddot{v}(t)^F_{1,x}$	$\ddot{v}(t)^F_{1,y}$	$\ddot{v}(t)^{G,C}_{1,x}$ and $\ddot{v}(t)^{G,C}_{1,y}$	$\ddot{v}(t)^S_{1,x}$	$\ddot{v}(t)^S_{1,y}$
Base	–	–	Base sensor	–	–
S-4					
Third floor	$\ddot{v}(t)^F_{3,x}$	$\ddot{v}(t)^F_{3,y}$	$\ddot{v}(t)^{C,G}_{3,x}$ and $\ddot{v}(t)^{G,C}_{3,y}$	$\ddot{v}(t)^S_{3,x}$	$\ddot{v}(t)^S_{3,y}$
Second floor	$\ddot{v}(t)^F_{2,x}$	$\ddot{v}(t)^F_{2,y}$	$\ddot{v}(t)^{G,C}_{2,x}$ and $\ddot{v}(t)^{G,C}_{2,y}$	$\ddot{v}(t)^S_{2,x}$	$\ddot{v}(t)^S_{2,y}$
First floor	$\ddot{v}(t)^F_{1,x}$	$\ddot{v}(t)^F_{1,y}$	$\ddot{v}(t)^{G,C}_{1,x}$ and $\ddot{v}(t)^{G,C}_{1,y}$	$\ddot{v}(t)^S_{1,x}$	$\ddot{v}(t)^S_{1,y}$
Base	–	–	Base sensor	–	–

To measure the drift inclination during the entire dynamic process, two bi-directional drift inclinometers were mounted at the third-floor level at the SS and FS of TU structures. The arrangement of these drift inclinometers is summarized in Table 3.12 (where $\theta(t)$ refers to drift inclination time history). The number in the subscript defines the floor number whereas the alphabet defines the response direction. The superscript refers to the corresponding edge of the structure.

A detailed arrangement of the experimental setup for a representative steel model is presented in Figure 3.17, where the arrangement of instruments can be seen along with the method of application of floor mass for floor eccentricity variation.

Table 3.12 Drift inclinometers and their arrangement in steel models

Floor no.	Flexible side (FS)	Stiff side (SS)
S-1 (bi-directional asymmetric)		
Third floor	$\theta(t)_{3,x}^{F}$	$\theta(t)_{3,x}^{S}$
S-1 (mono-symmetric)		
Third floor	$\theta(t)_{3,x}^{F}$	$\theta(t)_{3,x}^{S}$
S-2		
Third floor	$\theta(t)_{3,x}^{F}$	$\theta(t)_{3,x}^{S}$
S-3		
Third floor	$\theta(t)_{3,x}^{F}$	$\theta(t)_{3,x}^{S}$
S-4		
Third floor	$\theta(t)_{3,x}^{F}$	$\theta(t)_{3,x}^{S}$

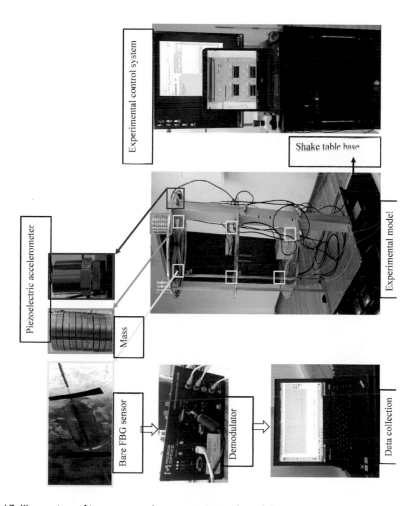

Figure 3.17 Illustration of instrumented asymmetric steel model

3.7   Input excitations

The experimental models were tested under various seismic excitations. Different El Centro earthquake records were used for concrete and steel models. It should be noted that the illustrated seismic components in Figures 3.18 and 3.19 are different because the presented seismic components correspond to different seismic events. Table 3.13 reports a detailed summary of the earthquake records used in this book.

The strategy for input seismic excitation for both RC frame-shear wall model and steel models is explained in the next subsections.

3.7.1   Input excitations for RC model

The direction of excitation was considered along the Y-direction (transverse direction) only because of the limited capacity of the shake table equipment, using two waveforms: white noise and the El Centro 1940 earthquake record. Before the test, white noise was used to identify the model frequency and mode. The input small amplitude white noise signal was used to convert the response time domain signal into Fourier transform to frequency domain signal, which can measure structural dynamic properties (including mode shape, damping ratio and natural frequency) at each stage. The El Centro (NS 1940) earthquake record was considered for seismic testing because of the fact that it is the most widely used high-frequency earthquake around the globe; therefore, the test result will be helpful for comparison of previously available data. Seismic inputs were progressively applied during the test, i.e progressively increasing acceleration amplitude of excitation; each level was increased with a PGA of 0.1g in order to obtain an elastic model structure at each stage until the plastic state was achieved. The progressive increase in the seismic excitation is illustrated in Figure 3.18 where a gradual increase of PGA = 0.1g at each testing level has been shown in the form of acceleration-displacement response for the El Centro ground motion along with the time history and velocity response. The input ground excitations have been represented as the acceleration-displacement spectrum so as to have a better idea with the damage investigation conducted in this research. Table 3.14 also represents the progressive increase in the ground motion at each test level.

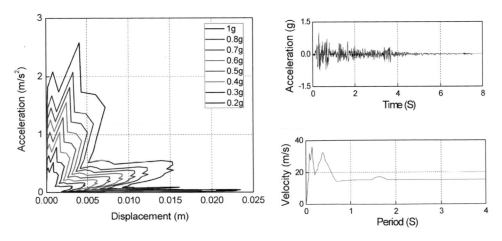

Figure 3.18 Seismic input for RC model

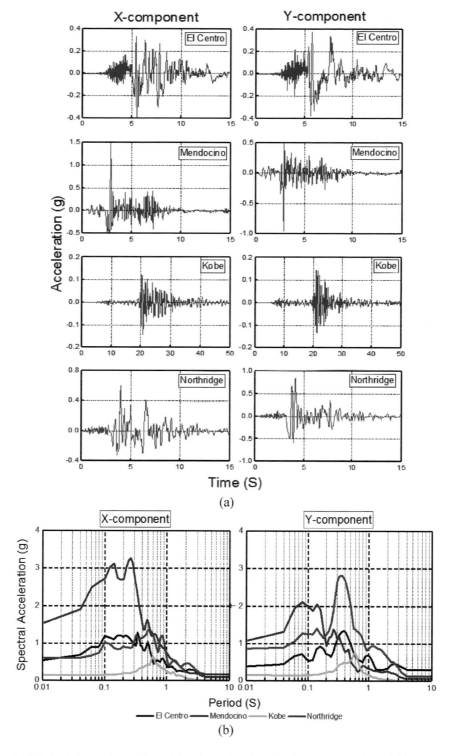

Figure 3.19 Seismic input for steel models: (a) acceleration time history records and (b) acceleration spectrums of the input excitations

Table 3.13 Summary of considered ground motion components

Earthquake	Station	Mechanism	Magnitude	Peak ground acceleration (g)		Soil condition
				X-component	Y-component	
Imperial Valley 19/05/40	El Centro Array #9	Strike-slip	6.95	0.1 to 1.0 (progressive increase)	–	Alluvium
Imperial Valley 15/10/79	USGS 952 El Centro Array #5	Strike-slip	6.53	0.4	0.39	Alluvium
Cape Mendocino 25/04/92	CSMIP 89005	Reverse	7.01	1.5	1.0	Rock
Kobe 16/01/95	HIK	Strike-slip	6.9	0.14	0.14	Sandstone
Northridge 17/01/94	County Hospital	Reverse	6.69	0.6	0.84	Alluvium

Table 3.14 Loading program for RC model

Test no.	Input excitation	Peak ground acceleration (g)
1	White noise	0.05
2	Imperial Valley/El Centro	0.2
3	White noise	0.05
4	Imperial Valley/El Centro	0.4
5	White noise	0.05
6	Imperial Valley/El Centro	0.5
7	White noise	0.05
8	Imperial Valley/El Centro	0.6
9	White noise	0.05
10	Imperial Valley/El Centro	0.7
11	White noise	0.05
12	Imperial Valley/El Centro	0.8
13	White noise	0.05
14	Imperial Valley/El Centro	1.0
15	White noise	0.05

3.7.2   Input excitations for steel models

Experimental testing was initially conducted on TB models followed by testing on various TU models. Before exposing the experimental models to bi-directional seismic excitations, the models were excited using a chirp wave to investigate the modal parameters of the structures under investigation. All the models were excited under four different seismic waves of various characteristics including near-fault and far-fault seismic waves with their dominant period of vibration closer to the estimated natural period of the structure. Considered ground motion excitations correspond to the El Centro, Mendocino, Kobe and Northridge

Table 3.15 Loading program for steel models

Test no.	Input excitation	Peak ground acceleration (g)	
		X-direction	Y-direction
1	Sine wave	0.05	0.05
2	Imperial Valley/El Centro 1940	0.40	0.39
3	Mendocino	1.50	1.0
4	Kobe	0.14	0.14
5	Northridge	0.60	0.84

earthquakes, where the Kobe earthquake represents a far-field earthquake and the rest of the earthquakes are representative of near-field earthquakes. The seismic excitation was not scaled down considering the nature of this work. A summary of the seismic loadings used for each steel model is reported in Table 3.15.

The time histories of the considered earthquakes in the orthogonal plan are illustrated in Figure 3.19a. In addition, the bi-directional elastic response spectrums of the considered seismic excitations are given in Figure 3.19b. The shake table incorporated for these tests has the bi-directional shaking vibration capacity of 1.0g in the horizontal plane. The experimental investigation was conducted in three different steps. In the first step, dynamic properties of all the steel models were evaluated. Then, symmetric models were tested for comparison purposes. Finally, detailed experimental testing on asymmetric structures was carried out.

3.8   Summary

Five types of models C-1 (RC model) and S-1 through S-4 (steel models along with their symmetric counterparts) were constructed and considered for the experimental evaluation. The RC model was seismically excited under a progressive increase of the seismic excitation to assess the actual physical deterioration in the RC model and to evaluate the behavior of the FS and SS considering the structure as stiffness eccentric structure. Models S-1 through S-4 were designed to represent a structural group from highly TS to highly TF. The purpose of these models is to simulate the damage behavior and to evaluate the global structural response under the interaction of various irregularities. This chapter gives a detailed picture of the construction of the experimental models, their geometric configurations, the pattern of irregularities, instrumentation and input seismic excitations.

Chapter 4

Damage response investigation in asymmetric structures

4.1 Introduction

This chapter provides a detailed investigation on the seismic damage response of asymmetric structures discussed in the previous chapters. It should be noted that the C-1 model experienced highly inelastic deformations under high seismic excitations, which eventually caused the transformation of the C-1 model from a state of elastic deformation to a state of highly inelastic deformation. In this book, these transformations are considered as the variation in the eccentricities of C-1 model at each experimental step. Damage in the form of plastic hinges occurred because of induced dynamic instability in the structure. For a broad range of asymmetric characteristics, this chapter provides contour plots of strain transition from the FS to the SS and from the SS to the FS. The monitored seismic responses in this chapter include the variation in the residual strains and dynamic strain time histories for the RC model, whereas for steel models, detailed contour plots of the monitored strain at the FS and SS are illustrated and analyzed.

This chapter is helpful in providing precise and detailed measurements of varying strain responses and effective estimation of damage response in asymmetric structures. Moreover, successful implementation of fiber Bragg grating (FBG) strain sensor is presented and applied in this experiment because of its enhanced sensitivity towards monitoring the structural dynamic strain response. The book also evaluates the suitability of FBG strain sensors in dynamic testing of asymmetric structures by comparing the damage response obtained through FBG sensors and the predictions developed from the dynamic properties for C-1 model, monitoring the progress in structural damage and predicting the cracks inside the structure.

4.2 Contribution of this chapter to knowledge

Previous literature related to FBG-based health monitoring of RC structures does not fully guide us about the damage response in building structures, and the main reason behind this is that the majority of FBG-based health monitoring studies are limited to bridges, dams, tunnels and other infrastructure facilities. Its applications are not as widespread to buildings as its applications to other infrastructure facilities. Moreover, previous research on FBG sensors provides an insight into the structural performance with apparently no physical damage to the structure. Estimating the damage response of a physically damaged structure is yet another complicated process and highly depends upon the type sensing technique and location of the sensors. It should be noted that despite the extensive use of FBG sensors in various infrastructure facilities, there is nearly no previous research on FBG sensor-based damage response in RC building structures. Therefore, there is a need to fill this research gap by detailed damage investigation in building structure using FBG sensors where a structure actually transforms from an elastic state to an inelastic state under the formation of cracks followed by progressive seismic excitations.

It should also be noted that damage investigation of RC buildings by traditional methods does not truly depict the building's damage response. The main reason behind this is that the usual practice for damage response monitoring in a shaking table test of a structure involves the use of traditional strain-based damage measurement methods where resistance strain gauges are used on the surface of concrete for strain measurement. Such methods are only useful in monitoring the external deformation variations. Since the external deformation of the structure cannot reflect the damage degree inside of the structure, accurate measurement of strain variation and effective prediction of damage inside the structure is one of the major objectives of this book. Moreover, liquid adhesives such as cyanoacrylate glue, RTP-801 adhesive or acrylate adhesive are used to attach these strain gauges onto the surface of structures. Since the strength of emulate concrete materials is smaller than the strength of these adhesives, they are not appropriate for attaching strain gauges to the external surface of the experimental models. These instant adhesives for fixed strain gauges can react with model material, and eventually they can influence the measurement accuracy. Besides, they are also vulnerable to having reduced insulation resistance under intense shaking as they are likely to get removed or displaced followed by the damage in the structure and eventually the accuracy of the measurement gets compromised. Some other key points that further enhance the novelty of this book are as follows:

1   As explained previously, the research on FBG-based damage response asymmetric buildings undergoing inelastic deformations is nearly nonexistent. Therefore, FBG strain sensors were used in this study as these sensors have the ability of periodical variation in the index of refraction of the optical fiber core. Because of its numerous advantages over other technologies and its appropriate features which include its embedding abilities, high sensitivity, electro-magnetic interference immunity and flexibility, its applications in successfully investigating the damage process are evaluated.

2   The existing data stock on experimental damage response of asymmetric structures is not significant; therefore, data collected in this research can be used for future evaluation of damage response, verification and development of computational models for prediction of damage behavior in asymmetric RC structures.

4.3   Fiber Bragg grating sensing principle

Careful handling and protective housing for the embedded FBG strain sensors are required when laying these sensors in concrete structures. A perfect bond between the concrete and protective housing is also required to be ensured for faithful monitoring of the structural strain through sensors. The widely used embedded FBG sensors are the ones packaged by capillary or steel tube. In some cases, the adhesion of a steel tube with low strength material does not cause sufficient deformation in the steel tube. Therefore, this phenomenon causes strain transfer loss due to decreased sensitivity of sensors. There is always a need for sensitivity enhancement of FBG strain sensors. The refractive index FBG strain sensors change cyclically along the axial direction of the fiber. The current principle of various sensors based on fiber Bragg gratings can be attributed to the measurement of the center wavelength λ_B of the Bragg grating, that is, by measuring the drift caused by the external disturbance. The measured parameters were obtained through Equation 4.1 and are related to the fiber grating length period Λ and effective refractive index η_{eff} of the fiber core.

$$\lambda_B = 2\eta_{eff}\Lambda \tag{4.1}$$

The famous principle of end bearing piles and friction widely used in civil infrastructure related works was utilized to increase the cohesive force between the measurand materials. The cohesive force and the surface area roughness between the measurand materials were increased by designing the cube shapes and mounting it at the supports. The mounting supports were fixed with grippers and both sides of the FBG sensors were set with the mounting supports in order to have a better transfer of cohesive force to the sensor. A steel tube was used for packaging of the bare FBG sensors located between the grippers. Epoxy resin was used to encapsulate the fiber in FBG sensors in the grippers. The grippers on the mounting support were installed by solder. The installed steel tube serves only as a protective housing and does not transfer strain to the sensor during the process of pouring the concrete. Also, grippers and mounting supports could freely slip through the steel case. The force due to cohesion between the material and mounting support causes the slip of grippers along the steel tube. This eventually yields in the deformation of bare FBG and grippers. The illustrations have been presented in Figures 4.1 and 4.2.

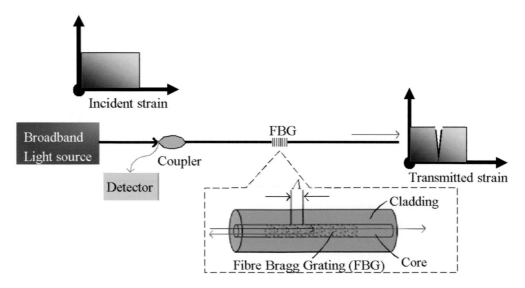

Figure 4.1 Schematic diagram of stainless steel tube-packaged FBG sensor

Figure 4.2 Stainless steel tube-packaged FBG sensor

4.4 Damage characteristics and its measurements

4.4.1 Physical damage characteristics of RC model

In this experiment, the structure is a frame-shear wall asymmetric structure. When the strain was monitored, the maximum strain corresponding to the axial tensile strength was about $\varepsilon_{0t} = 0.00015 - 0.0002$ respectively, usually taken $\varepsilon_0 = 0.00015$, that is $150\mu\varepsilon_0$. The strain response of the structure was monitored through installed sensors.

Figure 4.3 provides the strain history profiles of all the sensors that were installed in the structure for PGAs of 0.2g and 1.0g. The strain history profiles have been incorporated with a zoom response at peak points. The tensile strains correspond to the positive strain in strain history profiles while the compressive strains correspond to negative strain in the strain history profiles.

As can be seen in Figures 4.4 and 4.5, the maximum tensile strains observed by sensor #1, sensor #2, sensor #5, sensor #10, sensor #11 and sensor #12 reached the maximum limit when the maximum input ground acceleration was 0.2g. When the test structure was exposed to maximum input ground acceleration of 0.3g, the tensile strain monitored by sensor #3, sensor #8, sensor #9 and sensor #13 reached the maximum limit of ε_{0t}. The location of these sensors is between the frame bars of the two columns of the first story and the beam between the side column frames.

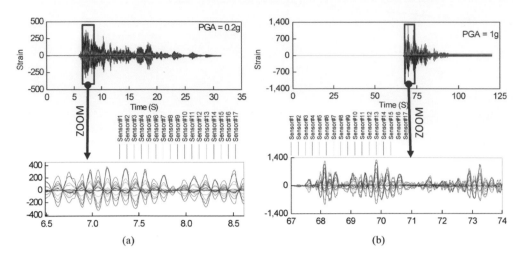

Figure 4.3 Recorded strain time history profiles of C-I model; recorded strain profiles corresponding to all FBG sensors for input ground motion of (a) 0.2g and (b) 1.0g

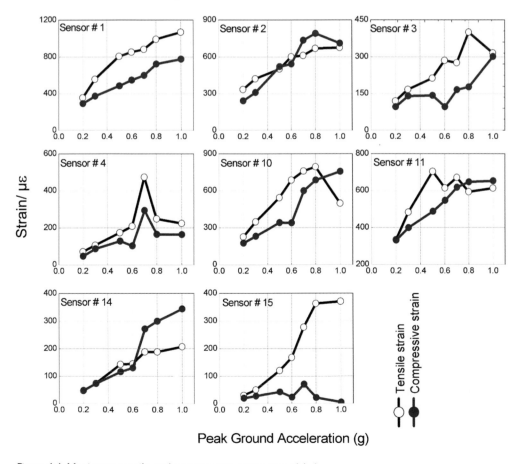

Figure 4.4 Maximum tensile and compressive strain at grid A-I

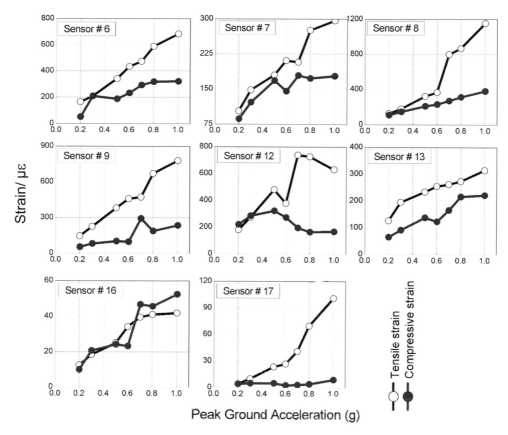

Figure 4.5 Maximum tensile and compressive strain at grid B-1, grid B-2 and shear wall

When the input ground motion was further increased to 0.5g, the strain recorded by all the sensors reached the maximum limit of tensile strain except the two sensors installed in the shear wall, sensor #16 and sensor #17, which did not reach the maximum limit of tensile strain. However, the other points were already in the limit state when a PGA of 0.5g was applied to the test structure. It can be seen in Figures 4.4 and 4.5 that when the input ground acceleration was further augmented from 0.5g to 0.6g, the strain in sensor #10 and sensor #11 became smaller due to the fact that sensor #10 and sensor #11 had a tension force. The edges of the two beams between the column frames were closer to these two sensors where the cracks were also present. The reason for the smaller strain was not due to the placement of sensors in the vicinity but due to the presence of cracks near the sensors. These cracks near the sensors released some part of the sensor strain; therefore, the augmentation of input ground motion yielded smaller tensile strain due to the release of part of the strain at the crack. When the PGA was further increased to 0.7g, some very interesting phenomena were observed. The maximum tensile strain of sensor #4 was almost doubled while the compressive strain was increased to nearly 3 times the compressive strain observed in the previous PGA. This shows the presence of a large crack near this region and also indicates that concrete was crushed at this point. This eventually confirms the formation of a large crack and crushing of concrete at 0.7g. It is interesting to note that the cracks here appear to cause the original concrete to bear the tensile load developed by the rebar and eventually lead to cracks in the concrete. After the concrete was crushed, the compressive strain induced in the concrete was developed and transferred from the rebar.

The maximum tensile strain of sensor #8 was also increased to double. The compressive strain did not have a significant change which indicates the fact that the concrete here was also cracked but was not crushed. When the column was pulled by the concrete, the tensile strain was developed by the rebar. Because of the compression, the concrete and the steel bars together were subjected to the compressive strain. When the peak value of ground motion was increased from 0.7g to 0.8g, the maximum tensile strain and compressive strain of sensor #4 changed to 1/2. This shows that the structure was in the plastic state, where the plastic deformation of steel formed a plastic hinge at the joint. The compressive strain of sensor #15 was negligible which confirms the fact that a plastic hinge was formed at this node as well.

4.4.2   Damage simulation in steel models

Figure 4.6 represents the assumption considered for local response investigation. Since the deformation concentration has been presented as strain contours in Figures 4.11–4.15, the ideal scenario for strain contours should have been a distributed pattern of FBG strain sensors all along the line from the FS to the SS of the structure. However, it has been assumed that the trend in the strain variation is uniform when the structure is composed of steel material. Therefore, bare FBG sensors were pasted only at the SS and FS. The obtained responses have been developed as strain contours to reflect the tensile and compressive deformation concentration at the SS and FS of the experimental models.

For monitoring of the strain responses, the pasted bare FBG strain sensors were glued onto the strategic locations to cover the interests of this book. The deployment of these sensors is presented in the Figures 3.15 and 3.16. The determined strategic locations for the deployment of the bare FBG sensors include the FS and SS of the asymmetric structures. The monitored strains of the FS and SS were calculated based on the shift in the wavelengths of the FBG sensors. Considering the sensitivity of the local strain response, the installation of FBG sensors demands very careful handling; therefore, bonding and protection of FBG sensors were carried out very carefully.

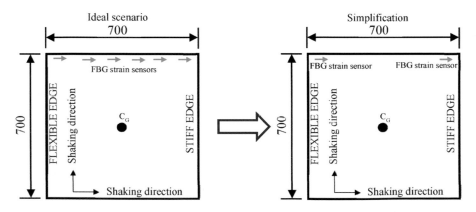

Figure 4.6 Simplification for the local response evaluation in S-1, S-2, S-3 and S-4 models

The understanding of the structural response, in particular as asymmetric structures under bi-directional seismic excitations are concerned, becomes crucial when seismic response comparison is considered at the FS and SS of the asymmetric structure. This is due to the complex nature of the local response transition from compressive deformation to tensile deformation and from tensile deformation to compressive deformation in the time domain when the asymmetric structure is exposed to bi-directional seismic excitations. The main goal here is to assess the behavior of local strain concentration at the FS and SS of asymmetric structures under various input seismic waves. Therefore, the behavior of the local seismic response is monitored using bare FBG sensors at strategic locations and for the ease of understanding strain concentration at the SS and FS are plotted together as strain contours.

The test results of steel models also reflect the effective pattern recognition system of deformation concentration mechanism and large distortions without appreciable damage in the time domain under varying eccentricities. Local strain response was monitored during the dynamic testing when the asymmetric structure experienced bi-directional seismic excitation. The strain measurements at the FS and SS of all floors under various asymmetries have been described in the previous chapter. Moreover, it is imperative to differentiate between the behaviors of the strain response at the two edges when the irregularity is varied in various cases of asymmetries.

The first set (case 1) corresponds to the investigation of the structural behavior in the reference state (Fig. 4.11). In the second step, local response investigations were conducted when the level of eccentricity was further aggravated by shifting the C_M away from the C_R. The focus was the assessment of strain concentration and correlation of the response between the SS and FS. Then, different levels of eccentricities simultaneously occurring at all floor levels are projected in terms of strain contour profiles corresponding to the SS and FS.

For all the models under investigation, sine wave excitations with low acceleration peak were used to evaluate the dynamic characteristics of the structure. Figures 4.11–4.15 illustrate the strain contour plots achieved from bare FBG strain sensors pasted onto the surface of steel models. The achieved strain history curves recorded at the SS and FS of TU systems, along with their counter TB system, describe the similar pattern of input seismic excitation. The areas to be monitored are schematically highlighted in Figure 3.17. Due to asymmetry in both shape and loading condition, the expected mechanical response also shows asymmetric behavior.

The contour plots under dynamic input for the illustrated strains are represented for the bi-directional Northridge earthquake only with the maximum vibration period of 0.5s in the X-direction and 0.35s in Y-direction. Plots corresponding to other excitations can be seen in Appendix B.

For the purpose of analysis, only a few cases have been illustrated in Figures 4.11–4.15. Detailed plots corresponding to other cases can be seen in Appendix B. The responses can also be useful in correlating the local response with global response. The contour plots in Figures 4.11–4.15 illustrate the FS vs SS contours of the local strain response in the time domain observed at each floor level. These contour plots are not only representative of the particular edge of interest but are representative of particular floor of interest as well. In the presented contour plots, positive and negative strain regions are the representative tensile and compressive strains correlated for the FS and SS at all floor levels. In each of the presented contour patterns, there are asymmetrical features where higher tensile regions extend either towards FS or SS. It is specifically these regions where cracks are most likely to occur and propagate. The asymmetric behavior of the structures at all floor levels is obvious in all the contour plots. It is interesting to note that the influence of location-specific eccentricity is more evident in these plots compared with the global responses which will be explained in later sections. It can be concluded that these plots are more sensitive to the location of eccentricity and therefore response corresponding to the floor of higher eccentricity has demonstrated higher tensile and compressive strains. Moreover, it can be seen that the floor containing higher eccentricity has demonstrated higher influence on the SS compared with the FS of the structure. However, tensile strain concentration is more important to understand for the reason to attribute its nature to the cracking in the structure.

4.5   Local deformation concentration at FS and SS

4.5.1   Local response in RC model

4.5.1.1   Elastic response at FS

The observations made in this section from strain history profiles do not consider the initial strain caused by the casting and pouring of concrete. The explanation about the initial strain consideration is explained in section 9. The FBG sensors labeled "sensor #14" and "sensor #15" with their location being provided in Figure 3.14 at the beam-column joint have been selected to analyze the damage characteristics and stability of the model under progressive input loading. Figure 4.7 illustrates the strain time history profiles in the elastic state for the input ground motions of 0.2g and 0.3g, respectively. It can be seen that the strains of both sensor #14 and sensor #15 increased gradually with the increase in the input ground motion. The structure remained in the elastic state at PGA of 0.3g and no visible cracks were noticed at this stage.

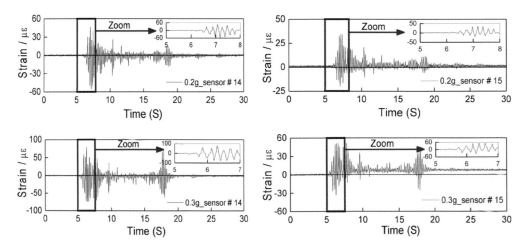

Figure 4.7 Elastic strain response at the FS of C-1 model

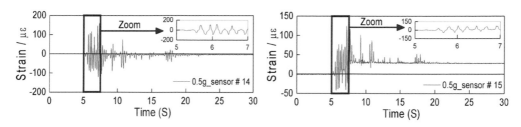

Figure 4.8 Strain response in the micro-cracking state at the FS of C-1 model

4.5.1.2 Micro-cracking response at FS

The state of internal micro-cracking was determined from the variation in the dynamic properties of the structure as structural frequencies started to fall significantly in first mode when the input ground acceleration was 0.5g. In Figure 4.8, strain profiles at the internal micro-cracking state are illustrated. It can be seen that strain response in sensor #15 shows a complete irrecoverable plastic deformation with significantly high residual strain in the tensile range. There is some amount of compressive stress present as well, which can be neglected as the strain response will gradually shift into the tensile range once the seismic loading is further increased.

4.5.1.3 Inelastic response at FS

As the input ground acceleration intensity increased from 0.5g to 1.0g, the shear wall frame structure started to transform to the inelastic state. Also, it was found that tensile strain was greater than the compressive strain in most strain time history profiles. Finally, it was discovered that compressive strain was found to be higher than tensile strain in the plastic state for sensor #14 and tensile strain was found to be higher than compressive strain in the plastic state for sensor #15 (Fig. 4.9). In the elastic state, both sensor #14 and sensor #15 reported fairly equal tensile and compressive strain peaks. These observations demonstrate damage scenarios from sensor to sensor.

4.5.1.4 Elastic and inelastic response at SS

In comparison with the structural behavior at the FS of the structure, this section provides an insight into the structural response at the SS of the structure where it can be seen that structural deformation at the SS is negligible in all loading states. Figure 4.10 demonstrates the strain response obtained from sensor #16, located at the first-floor level in the shear wall. The demonstrated strain profiles prove the fact that the SS of the structure hasn't experienced severe deformations and thus no apparent damage was discovered at the SS of the structure. Besides, it can be seen that both compressive and tensile strains remained fairly equal in all loading states. Two sensors (sensor #16 and sensor #17) were installed in the shear wall at the first- and second-floor levels. However, for reasons of compactness, results from sensor #16 have been illustrated here. For a better picture of structural response at the SS, Figure 4.10 needs to be observed in combination with Figures 4.4–4.5, where absolute tensile and compressive strains have been demonstrated for both sensor #16 and sensor #17.

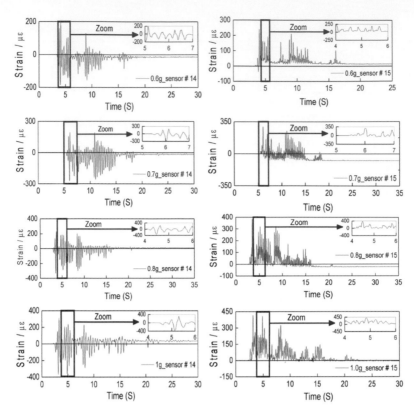

Figure 4.9 Inelastic strain response at the FS of C-1 model

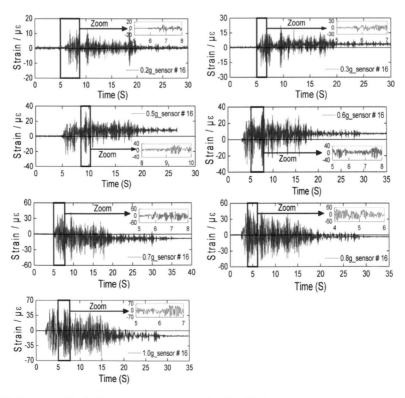

Figure 4.10 Elastic and inelastic strain response at the SS of C-1 model

4.5.2   Damage simulation in steel models

In Figure 4.11, case 1 refers to the reference state of the structure. It can be seen that for the S-1 structure (TS-square-shape structure), simultaneous tensile state at the first-floor level for both the FS and SS is negligible. However, at intermediate and top roof levels

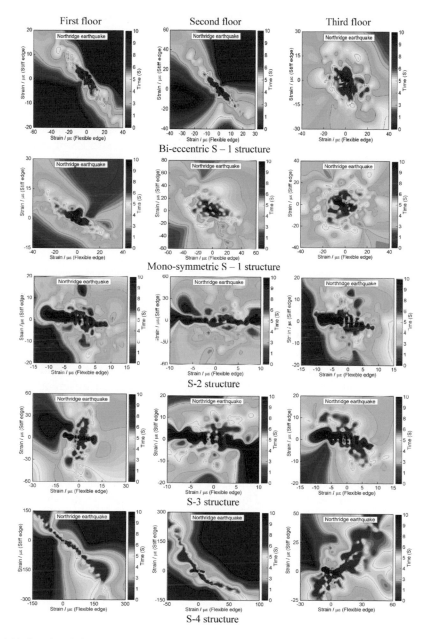

Figure 4.11 Simulated damage response at flexible and stiff edge for case 1 under Northridge earthquake: first column corresponds to the response at the first-floor level; second column corresponds to the response at the second-floor level; third column corresponds to the response at the third-floor level

for the same structural type, the pure tensile and compressive regions tend to spread with simultaneous occurrence of tensile as well as compressive response at the FS and SS. In terms of absolute compressive and tensile strain response, it can be seen that the SS of the structure demonstrates lower or negligible degree of tensile strain at the first-floor level. However, this is not true for the top roof level where FS tends to have higher compressive and tensile strains compared with the strain response at the SS of the structure. For the S-2 structure (TF square-shape structure) with stiffness and strength eccentricity, it can be seen that neither the first-floor nor top roof level has experienced simultaneous compressive strains. In terms of tensile strains, both the floors experienced entirely similar simultaneous tensile strains at the FS and SS of the structure. More-over, it can be seen that the intermediate floor experienced higher tensile strains at the SS of the structure compared with FS. This unusual response occurred because of the contribution of second mode dominance. For the S-3 structure, it can be seen that SS experienced higher compressive and tensile strains compared with FS of the structure. Moreover, the influence of local deformation is higher compared with the intermediate and top roof levels. It should also be noted that the tensile deformation is higher at the SS compared with the FS of the structure. In terms of compressive strains, the first floor has experienced negligible compressive strain at the first floor compared with the top floor of the structure regardless of the influence of FS and (or) SS of the structure. For the S-4 structure with stiffness and strength eccentricity, it can be observed that lower floors are expected to experience higher damage compared with the upper floors. How-ever, in terms of simultaneous compressive and tensile deformations at the FS and SS of the structure, it can be seen that first-floor and intermediate level demonstrate negligible influence. Compared with this, the top roof level experienced simultaneous influence of compressive and tensile strains though the intensity is low. In terms of absolute tensile and compressive deformation, an abnormal pattern is observed at all floor levels. The first floor is sensitive towards higher tensile deformations at the FS whereas in terms of compressive deformations, SS appears more sensitive compared with the FS of the structure. The second floor experienced similar response with a difference that now the behavior of the FS and SS has interchanged. At the top roof level, the intensity of the absolute compressive and tensile deformations has been reduced quite significantly. Both the SS and FS experienced equal tensile and compressive deformations at the top roof level.

Of these contour plots, the positive value means tension while the negative value means compression. It can be seen that the strains of both the SS and the FS increase gradu-ally along the height of the structure. This phenomenon demonstrates that the lower floor remains in the perfectly elastic state, as the torsional vibrations tend to shift its influence at the higher floor levels. Overall, the tensile strains are greater than the compressive strains on the SS while the compressive strains are greater than the tensile strain at the FS of the first-floor level. These phenomena demonstrate that in this state of the structure, SS will deform towards FS. The following conclusions can be extracted from the monitored strain responses:

- The direction of deformation for S-1 structures can be from the FS to the SS at all floor levels. The first floor demonstrates equally negligible compressive and tensile state whereas top roof demonstrates higher tensile state at the FS. In terms of tensile strains, the FS of the structure is sensitive towards tensile deformations. In general, for this particular case, the top roof level is expected to experience higher amount of local damage as the top roof level is sensitive to tensile strains with the pattern of deformation being from the FS to the SS of the structure.

• S-2 structure experienced negligible compressive strains at first-floor and top roof levels. However, higher tensile and compressive strains have been monitored at the SS of the intermediate floor under the influence of second mode. Moreover, it can be said that in this particular case, the intermediate floor is likely to experience higher damage compared with other adjacent floors. The deformation will travel from SS towards FS of the structure.

• As opposed to the similar case of the S-1 and S-2 structures, the S-3 structure has demonstrated equally higher compressive and tensile strains at the first-floor level and negligible compressive and tensile strains at the intermediate and top roof levels. Moreover, the FS experienced lower tensile strains compared with SS of the structure which indicates that deformation at the first-floor level will travel from SS towards FS.

• S-4 structure demonstrated quite abnormal response at all floor levels. The first floor experienced higher absolute tensile deformations at the FS of the first-floor level compared with the SS of the structure whereas the SS experienced higher compressive deformations compared with the FS of the structure. This eventually means that the deformation will travel from the FS to the SS of the structure at the first-floor level. At the top roof level, both the FS and SS experienced equal tensile and compressive deformations.

In Figure 4.12, case 4 refers to the state when the static eccentricity at the top roof level is increased while the intermediate and first-floor levels remain in the reference state of stiffness eccentricity as has been described in detail in Table 3.1. It can be seen that the S-1 structure has experienced lower tensile and compressive deformations but with similar patterns of deformation compared with case 1. For the S-2 structure, it can be seen that no major change in the deformation pattern occurred compared with case 1. The intermediate floor appears to be more sensitive towards damage response despite the fact that the structure under investigation is now IRI. Major tensile and compressive deformations occurred at the SS of the second floor which indicates that the direction of deformation is from the SS to the FS of the structure. First-floor and top roof levels were monitored with a similar pattern of deformation at both the FS and SS of the structure. However, in terms of absolute tensile deformations, the SS of the structure remained under higher tensile deformations. Hence, looking at the overall influence at all floor levels of S-2 structure, the SS of the structure had higher tensile deformations compared with the FS of the structure. This in turn leads to the conclusion that the tensile deformation pattern in this particular case is from the SS towards the FS of the structure at all floor levels. For S-3 structures, similar deformation patterns were observed as were monitored in case 1. However, a slight difference at the intermediate and top roof levels was observed where it can be seen that there is no appreciable difference between the FS and SS response. For the S-4 structure,

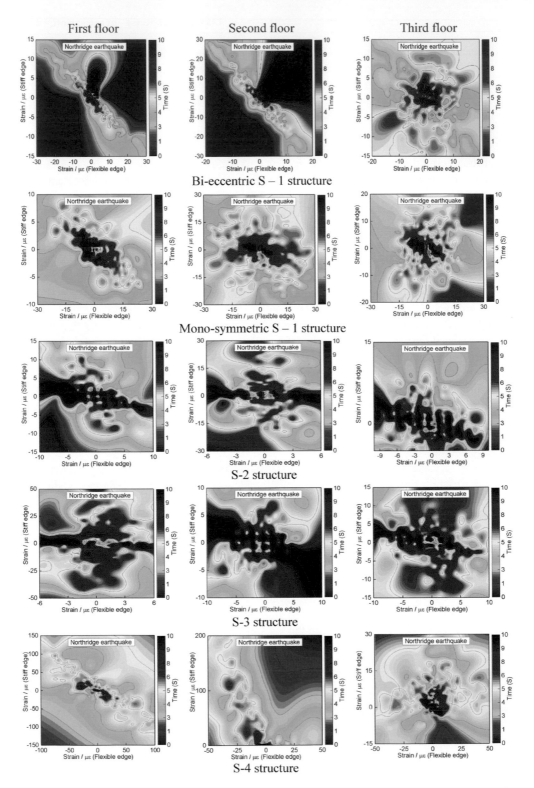

Figure 4.12 Simulated damage response at flexible and stiff edge for case 4 under Northridge earthquake: first column corresponds to the response at first-floor level; second column corresponds to the response at second-floor level; third column corresponds to the response at third-floor level

as expected, a similar deformation pattern was observed as was monitored for case 1 of the S-4 structure.

- The deformation pattern in the S-1 structure remained unaffected compared with case 1 regardless of the fact that the structure in this case is IRI whereas the structure in case 1 was RI. This indicates that for these kinds of structures, the top roof level eccentricities have the least influence on the structural local response. Moreover, deformations in the IRI state (case 4) have decreased enormously compared with the RI state (reference state or case 1).
- The S-2 structure experienced negligible compressive strains at first-floor and top roof levels similar to the deformation pattern in case 1. However, higher tensile and compressive strains have been monitored at the SS of the intermediate floor under the influence of second mode. Moreover, it can be said that in this particular case, the intermediate floor is likely to experience higher damage compared with other adjacent floors. The deformation will travel from the SS towards the FS of the structure. Furthermore, deformations in the IRI state (case 4) of S-2 structures have decreased enormously compared with the RI state of S-2 structures.
- For S-3 structures, similar deformation patterns were observed as in case 1, with a slight difference. In this case, no appreciable difference between the FS and SS response was observed at intermediate and top roof levels. Moreover, deformations in this case slightly decreased compared with the reference state of the structure.
- S-4 structures demonstrated a similar response as was observed in case 1. However, in this case, no appreciable reduction in the deformation was observed.
- In general, the deformation pattern of both the RI and IRI states of irregularity remained the same and the variation in the eccentricities at top roof level has caused the least influence on the deformation pattern of the structure. The overall deformations were decreased with an increase in the eccentricity at top roof level for TF structure. However, as the flexibility in the structure increased, this reduction in deformation decreased. For highly TF structures, no appreciable change in the deformation was observed.

In Figure 4.13, case 5 refers to the state when the static eccentricity at all floors of the structure is increased simultaneously as has been described in detail in Table 3.1. It can be seen that steel models experienced similar deformation patterns as were monitored in case 1. The deformations have slightly increased but overall the deformation pattern remained unchanged. This happened because this case is similar to the case of stiffness eccentricity (reference state or case 1) but with higher magnitude of static eccentricities at all floor levels. Therefore, it can be concluded that:

- Deformation patterns remain unchanged regardless of the uniform increase in the static eccentricity at all floor levels. However, the higher static eccentricities have slightly increased the magnitude of eccentricity at the first-story level for TS structures and at the first and intermediate story level for TF structures. This in turn leads to the conclusion that in TS structures, lower floors are sensitive to higher deformations compared with top roof level whereas in TF structures, this influence is transmitted to the higher floors simultaneously.

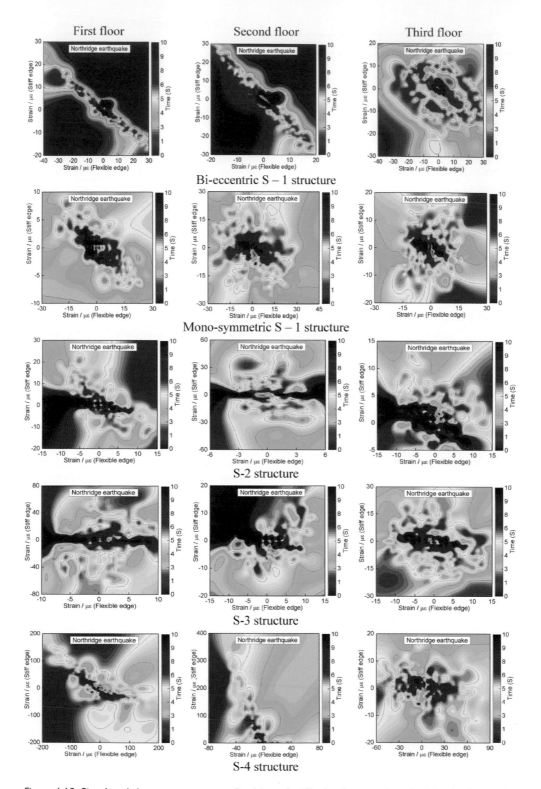

Figure 4.13 Simulated damage response at flexible and stiff edge for case 5 under Northridge earthquake: first column corresponds to the response at first-floor level; second column corresponds to the response at second-floor level; third column corresponds to the response at third-floor level

In Figure 4.14, case 6 refers to the state when the C_M and C_S are converged at one single point but dislocated from the G_C of the structure. It can be seen that this case has a major influence on the deformations compared with the previous cases of investigation. For S-2 structures, it can be seen that intermediate and top roof levels are under higher tensile and

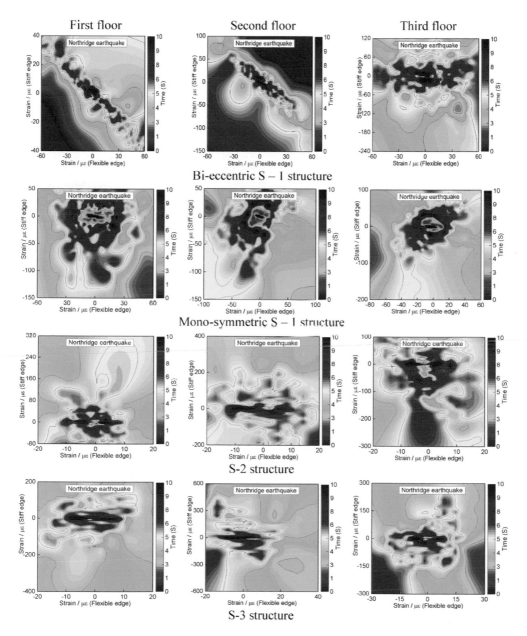

Figure 4.14 Simulated damage response at flexible and stiff edge for case 6 under Northridge earthquake: first column corresponds to the response at first-floor level; second column corresponds to the response at second-floor level; third column corresponds to the response at third-floor level

compressive deformations whereas the first floor remains unaffected. Moreover, higher tensile and compressive deformations are SS compared with the deformations at the flexible zone. This is because of the presence of heavy mass near the SS which was actually placed to dislocate the C_M. This has eventually influenced the deformations at the FS of the structure. However, this influence is entirely negligible at the first-floor level. It can be said that intermediate and higher level floors are sensitive to the damage response for S-2 structures under case 6. For S-2 structures, it can be seen that lower order floors experienced high tensile and compressive deformations compared with the top roof level. Moreover, the deformation pattern is from the SS to the FS of the structure. The reason for higher deformations at the SS is presence of heavy mass near the SS of the structure and therefore, it can be said that damage response is highly influenced under the location-specific scenarios. For S-3 structures, it can be seen that the intermediate floor experienced higher tensile and compressive deformations compared with the first-floor and top roof levels. For this particular case, S-4 structures have not been analyzed due to the disruption of the data.

- The deformation pattern in S-1 structures under case 6 significantly changed compared with the previous cases. Higher order floors are more sensitive to damage compared with lower order floors. Despite the significant change in the pattern of eccentricity, first-floor response remained approximately unchanged.
- Unlike S-1 structures, S-2 structures experienced higher tensile and compressive deformation on the lower order floors compared with the top roof level.
- S-3 structures experienced higher tensile and compressive deformations at the intermediate story level compared with the first-story level and top roof level. In terms of absolute compressive and tensile deformations, the SS experienced higher demands compared with the FS of the structure due to the presence of heavy mass near the SS.

In Figure 4.15, case 7 refers to the state when the vertical mass eccentricity is at the intermediate floor level whereas the first-floor and top roof levels remained in the reference state of stiffness eccentricity. This case has been analyzed to determine the location-specific influence on the local deformation demands. For S-2 structures, it can be seen that the FS is more sensitive towards compressive and tensile deformations as compared to the SS of the structure except the intermediate floor where relatively similar response was observed at both the FS and SS. For S-2 structures, the influence of location-specific eccentricity is clearly observable. It can further be observed that the floor corresponding to the vertical mass eccentricity has demonstrated higher compressive and tensile deformations at the SS of the structure compared with the FS. Intermediate and top roof floors have demonstrated similar response as was observed in case 1. However, more interestingly, the influence of seismic response in case of vertical mass eccentricity is location specific and has not transmitted its influence to the adjacent floors. Moreover, the sensitivity of the damage is limited to the SS of the structure and deformation flow is from the SS towards the FS. The main reason for the unusual behavior at the intermediate floor is the floor stiffness. For S-2 structures, the height of the intermediate floor is 1.5 times higher than the adjacent lower and upper floors. Therefore, the intermediate floor in this case serves not only the weak but soft floor as well. For S-3 structures, location-specific influence can be observed at the intermediate floor and it can be seen that the intermediate floor experienced higher compressive and tensile deformations

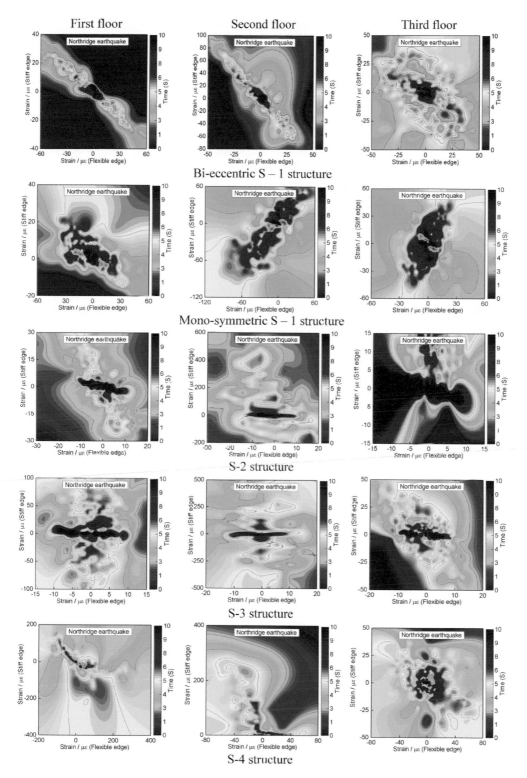

Figure 4.15 Simulated damage response at flexible and stiff edge for case 8 under Northridge earthquake: first column corresponds to the response at first-floor level; second column corresponds to the response at second-floor level; third column corresponds to the response at third-floor level

at the SS compared with the FS of the structure. Moreover, the influence of the vertical mass eccentricity is not only limited to the corresponding floor but has been transmitted to the adjacent lower floor. However, it should be noted that tensile and compressive deformations remained dominant at the SS throughout the height of the structure. For S-4 structures, a highly abnormal response has been observed due to the flexibility of the structure. Note that the intermediate floor in this case is the weak and soft floor. However, the location-specific influence not only affected the seismic response at the corresponding floor of eccentricity but has transmitted its influence to the adjacent lower floor. Moreover, it can be seen that the response at the SS of the intermediate floor is entirely tensile with no influence of compressive deformation. This influence can be used as an important tool for the damage prediction in such kinds of structures. Besides, the top roof level demonstrated similar behavior as was observed in the reference state of the structure.

- Unlike other cases, in the deformation pattern in the S-1 structures of case 6, the FS is more sensitive to higher tensile and compressive deformations compared with the SS of the structure except the intermediate floor where relatively similar observations can be seen for both the FS and SS of the structure. The deformation in this case is likely to travel from the FS towards the SS of the structure.
- For S-2 structures, the influence of location-specific eccentricity is dominant and the deformation pattern has been captured to be flowing from the SS towards the FS of the structure. Moreover, this influence is only limited to the floor corresponding to the higher vertical mass eccentricity. The first-floor and top roof levels remained unaffected in this particular case and demonstrated similar observations as were observed for the structure in its reference state. The reason behind this unusual behavior of the intermediate floor is because of its characteristics of being weak floor and soft floor. However, the weak floor and soft floor influence has only been seen under the influence of vertical mass eccentricity at the corresponding floor.
- For S-3 structures, location-specific eccentricity has not only influenced the corresponding floor but has transmitted its influence to the adjacent lower floor. The deformation travelled from the SS towards the FS at all floor levels. However, this influence is negligible in terms of compressive deformations at the SS of the structure.
- For S-4 structures, the deformation patterns have been greatly influenced not only at the floor of vertical mass eccentricity but the influence has been transmitted to the adjacent lower floor as well. The SS of the intermediate floor remained completely under tensile deformations. This pattern is an important parameter to predict damage response in such kind of structures. It should be noted from the general observation and results of the experimental investigation that in TF structures, the FS of the structure is the most affected region; however, when the vertical mass eccentricity is present at a floor, the response is greatly reversed and the deformation pattern appears to be traveling from the SS towards the FS rather than the FS towards the SS.
- In terms of local deformation response, location-specific eccentricities transmit its influence to the adjacent lower floor.

A detailed investigation corresponding to each type of structure, each case of eccentricity and under various seismic excitations is illustrated in Appendix B. For case 2, case 3, case 7 and case 9 of the steel models under Northridge earthquake and for all other cases under other seismic excitations, Appendix B can be used.

4.6   Damage investigation in terms of residual strains in RC model

Figures 4.16–4.17 illustrate the typical graphs of residual strains reported by all the sensors. The neutral axis is the term in material mechanics where the strain monitored by the strain sensor is defined as a "neutral axis" under a single ground motion. At the end of the vibration if the strain is consistent with the initial strain, that is known as unchanged neutral axis. If the strain at the end of the vibration is higher than the initial strain or below the initial strain, that is defined as neutral axis up and neutral axis down. Also, if the neutral axis has been declining under the action of several earthquakes, it indicates the presence of cracks. The location of these cracks would be in the vicinity of the sensors because the emergence of cracks would cause the decline of the neutral axis due to the release of tensile strain acting on the sensor (Fig. 4.21). The frequent movement of neutral axis under the action of ground motion indicates the crack near the sensor region. The tensile strain in that crack part is termed as residual strain. The concept of neutral axis was used to determine the residual strains corresponding to each of the installed sensors.

Considering the experimental results, these terms were applied in making the assessment. Also, a situation in which the neutral axis first moves up and then moves down is very complex in nature and therefore, it needs to assessed in combination with Figures 4.4–4.5 for clear understanding of the deformation behavior of the structure. The two described situations could form when the structure changes its states from elastic to plastic, forming a plastic hinge near the joint, or it can be formed when the reinforced elastically deformed structure restores a portion of the plastic deformation under the action of ground motion. This analysis has been made only for the sensors installed in beams and columns and does not consider the shear walls as the installed sensors in the shear wall did not record significant change in the strain due to higher stiffness of the wall. From Figures 4.16–4.17, it can be said that the neutral axis moved up for sensor #2, sensor #3, sensor #6, sensor #8, sensor #9, sensor #12 and sensor #13 with the first obvious upward shift when the ground motion inputs were 0.4g, 0.4g, 0.4g, 0.6g, 0.5g, 0.5g and 0.5g respectively. The neutral axis shifts with the order of neutral axis up and neutral axis down were observed for sensor #7 and sensor #15. For sensor #7 the neutral axis moved when the input ground motion was 0.3g and for the same sensor the neutral axis moved down when the input ground motion was 0.6g respectively. For sensor #15, the neutral axis moved up when the input ground acceleration was 0.4g and for the same sensor the neutral axis moved down when the input ground acceleration was 0.7g respectively. The neutral axis shifts with the order of neutral axis down and neutral axis up were observed for sensor #11 and sensor #15. For sensor #11, the decline of the neutral axis was observed when seismic excitation was 0.5g and for the same sensor the upward shift was observed when seismic excitation was 1.0g respectively. For sensor #14, the neutral axis drop was observed when seismic excitation was 0.4g and for the same sensor, the upward neutral axis shift was reported by the sensor when seismic excitation was 0.7g respectively.

As explained, assuming the pre-test strain of the sensor as the initial strain, the difference between post-test strain and the initial strain was termed as residual strain of the sensor. The residual strain reflects the redistribution of stress field inside the structure. Figures 4.16–4.17 illustrate that there was negligible residual strain in the installed sensors before the input ground motion was 0.3g and when the structure was completely

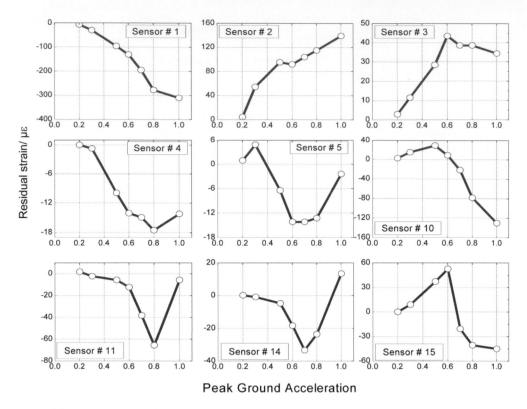

Figure 4.16 Residual strain under progressive loading at grid A-1

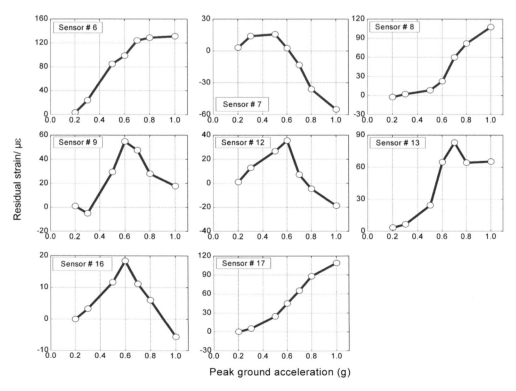

Figure 4.17 Residual strain under progressive loading at grid B-1, grid B-2 and shear wall

in the elastic state. The residual strains of FBG sensors began to increase remarkably when the input ground acceleration started to increase from 0.5g. This demonstrates the fact that initial micro-cracks appeared inside the structure due to the structure's irreversible elasto-plastic deformation. Besides this, the residual strain in all the sensors was found to be significantly large except sensor #16 and sensor #17. It shows that the shear wall region was the strongest zone of the structure and because of the high stiffness of the member, minimal relative strain changes were observed. The results also demonstrate that the sudden change in the residual strain at the FS in most of the profiles in Figures 4.16–4.17 is either due to the initiation of crack or crack travel through the experimental model. Therefore, the damage process in the experimental model could fruitfully be investigated using FBG strain sensors as its monitoring stands well with the observations.

In addition to the movement of the neutral axis of strain, there were special variations observed which were analyzed using Figures 4.7–4.9. No significant compressive strain (negative strain) was observed in sensor #15. In the vicinity of this sensor near the beam-column joint (Fig. 4.20d), the frame had a tensile failure, with failure being particularly near the ends of the beam-column joint. This tensile failure in the concrete was entirely developed by the rebars. Since the tension in the rebar caused a great plastic deformation and the concrete beam was under compression, plastic deformation didn't recover. In Figure 4.7, the strain time history profiles for sensor #15 under the input ground accelerations of 0.6g, 0.7g, 0.8g and 1.0g, respectively, indicate that the decline of the neutral axis with negative strain doesn't have an actual contribution towards the damage in the structure.

In Figures 4.16 and 4.17, non-uniform amplification in the residual strain mainly occurred in the micro-cracking state and inelastic state. In the elastic state, residual strain was negligible, even close to zero in some cases and therefore, no obvious difference in the residual strain was monitored in the elastic state. However, in the micro-cracking state and inelastic state, cracks appeared inside the concrete because of the irreversible elasto-plastic deformation, which further propagated with the progressive increase in the input excitation. Because of the formation of the cracks in the vicinity of the FBG strain sensors, some part of the strain released through the cracks and caused the redistribution of the stress field near the crack location. Moreover, the experimental model under consideration is an asymmetric reinforced concrete model with in-plan stiffness eccentricity. Under uni-directional seismic excitation acting perpendicular to the planar eccentricity, the flexible edge of the structure experienced non-uniform yielding, which also affected the strain response at the flexible edge and caused non-uniform increment in the residual strain. It is noteworthy to mention that Chapter 5 as a whole is an explanation of the cause of non-uniform increment in the residual strain, where a detailed argument is established on such effects along with the abnormal trend of the residual strain in the perimeter column of the asymmetric structure.

Since this experiment is based on the damaged process in the structure, when the ground motion input reached a certain peak value, the beam and column bars produced some residual strain in response to the input excitation. The FBG strain sensor provided the residual strain on the reinforcing bars by comparing the change of the wavelength before and after the change. The change in the wavelengths before and after the seismic action has been illustrated in Figures 4.18–4.19. Since this experiment was an indoor experiment, the influence of temperature was eliminated.

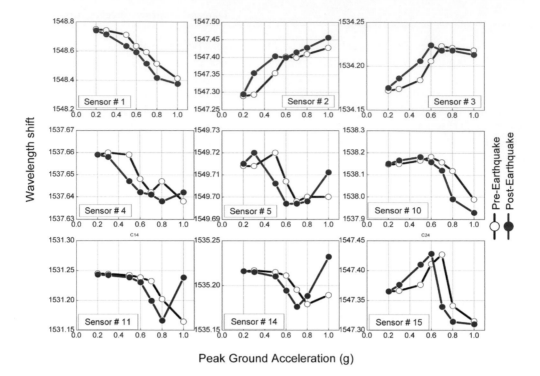

Figure 4.18 Wavelength shift under progressive loading at grid A-1

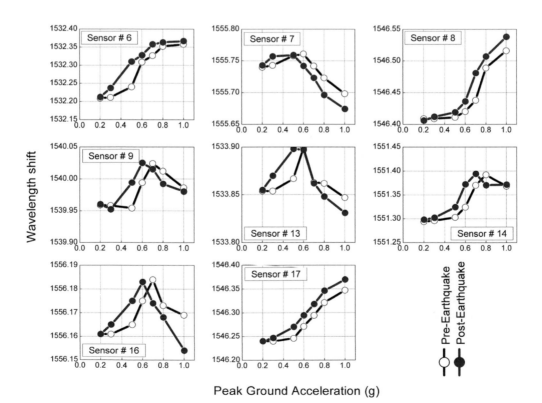

Figure 4.19 Wavelength shift under progressive loading at grid B-1, grid B-2 and shear wall

4.6.1   Initial strain consideration

The wavelengths of FBG sensors have been provided to consider the initial strain (shrinkage strain). The provided strain profiles have been monitored without considering initial strain due to curing and pouring. For more reliable results and better comparisons, these wavelengths can be used in correspondence with the temperature to obtain the initial strain of the overall structure before the test.

If T_p is the pouring temperature and T_t is the test temperature, then change in the temperature can be written as:

$$\Delta T = T_t - T_p \tag{4.2}$$

Also, if the center wavelength of the FBG during pouring is λ_p and center wavelength of the FBG during the test is λ_t then change in the wavelength $\Delta\lambda_B$ can be expressed as:

$$\Delta\lambda_B = \lambda_t - \lambda_p \tag{4.3}$$

where

$$\Delta\lambda_B = \alpha_\varepsilon \varepsilon_{ini} + \alpha_T \Delta T \tag{4.4}$$

In Equation 4.4, α_ε is the strain sensitivity co-efficient of the FBG sensor, ε_{ini} is the initial strain and α_T is the temperature sensitivity co-efficient of the FBG sensor. Simplifying Equation 4.4, the initial strain can be obtained as:

$$\varepsilon_{ini} = [(\lambda_t - \lambda_p) - \alpha_T (T_t - T_p)] / (\alpha_\varepsilon) \tag{4.5}$$

Equation 4.5 can be used to calculate the initial strain. For sensor #16, considering the pouring temperature as 25 degrees Celsius, test temperature as 5 degrees Celsius, center wavelength of the FBG sensor before pouring as 1547.837nm, center wavelength of the FBG sensor before the test as 1547.362nm, temperature sensitivity co-efficient as 0.027nm/°C and strain sensitivity co-efficient as 1.2×10^{-3}nm/$\mu\varepsilon$, the initial strain turns out to be 54.5$\mu\varepsilon$.

By considering the influence of the initial strain calculated in Equation 4.5, more reliable analytical data can be demonstrated for the analysis of either yielding of steel bars or ultimate tensile strain of steel bars.

4.6.2   Discussion on the formation of plastic hinges in RC model

The peak design acceleration was progressively increased in regards to the recommended intensity-PGA relationship (Xin et al., 2018; Du et al., 2017; GB17742, 2008) as described in Tables 4.1–4.3 for damage states of asymmetric structure at ultimate damage states. Since macro-seismic intensity is a direct measure of building damage, building stock require higher PGA for damage evaluation than were actually considered at the development stage of GB17742 (2008).

The seismic intensity in the 2008 Wenchuan earthquake was as high as XI according to the isoseismic map (Chen and Wang, 2010; Zhao et al., 2009). The reason for highlighting the damage and the corresponding higher intensity of this earthquake is because earthquake

Table 4.1 Recommended intensity-PGA relationship by the building code of China for a return period of 475 years

Intensity	VI	VII	VIII	IX	X
PGA (g)	0.05–0.09	0.09–0.18	0.18–0.35	0.35–0.7	0.7–1.4

Table 4.2 Recommended intensity-PGA relationship in China (Du *et al.*, 2017)

Intensity	VI	VII	VIII	IX
PGA (g)	0.05–0.12	0.09–0.18	0.22–0.41	0.41–0.75

Table 4.3 Recommended intensity-PGA relationship in China (Xin *et al.*, 2018)

Intensity	VI	VII	VIII	IX	X
PGA (g)	0.06–0.14	0.12–0.25	0.21–0.43	0.36–0.73	0.58–1.25

predictions did not help minimize the human and economic losses caused by major earthquakes in the world. It is the risk mitigation and redundant safety factors in the design process that could have effectively minimized the losses. Considering the fact that the evaluated structures in this study are asymmetric structures with various seismic response uncertainties described in Chapter 7, this book adopted a progressive increase in the input seismic excitation up to 1.0g, which is in line with the mentioned PGAs in Tables 4.1–4.3 for high intensity earthquakes.

The observed damage states in conjunction with the earthquake hazard levels can be summarized as follows:

1 Under minor seismic input, the entire behavior of the asymmetric structure was elastic. The maximum compressive strains in the moment resisting frame elements were less than 2/3 of the maximum compressive strength of concrete. At this stage, no significant nonlinear responses were observed in columns, as the compressive strain in the concrete was less than 2/3 of its compressive strength.

2 Under moderate seismic input, minor change in the dynamic properties of the structure was observed as reported in Table 4.4, which indicates the presence of micro-cracking in the structural elements. The initial micro-cracks were determined from the dynamic properties of the structure; after the structure was exposed to a ground acceleration of 0.5g. However, no visible cracks were noticed at this stage. The presence of micro-cracks was estimated from the variation in frequency of the structure.

3 Under rare seismic input, the overall behavior of the structure transformed into the inelastic state due to the irreversible plastic deformation. Residual strains at various locations on the flexible edge of the structure amplified with non-uniform increment due to stress redistribution near the crack location. Moreover, cracks started to become wider in the upper part of the frame near beam-column joints. This indicated that the

concrete stress exceeded the tensile strength of concrete at this loading state. However, at this stage, reinforcement bars did not yield.

4   Under progressive increase in the rare seismic input, various locations on the flexible edge of the structure experienced severe damage. At this loading state, the cracks started to become wider near beam-column joints and eventually resulted in the development of plasticity in the reinforcement. The corner beam in the transverse direction started forming bigger cracks near the beam-column joints while the middle columns in the longitudinal direction started forming bigger cracks both at the top and bottom edges. Damage to the moment resisting frame elements further aggravated with increase in the PGA of rare seismic event. In addition, plastic hinges started to develop near the beam-column ends at this level. When the input seismic excitation reached a peak value of 1.0g, cracks continued to increase and plastic hinges developed near the column ends.

In this research, it was discovered that the development of plastic hinges in the test structure was a progressive damage action which started from the structure's base and then slowly propagated towards the upward direction. It also started from the beams and then slowly and progressively propagated towards columns. The development of plastic hinges around the column and flexural damage at beam-column joints speeded up the damage process in the columns at the adjacent stories. As a consequence, the adjacent stories started to develop plastic hinges causing the formation of plastic hinges progressively. Therefore, it can be claimed that the initiation of structural damage started from the weak point of the structure because of lack of structural capacity following adjacent structural component's damage because of inertial forces being in the lateral direction. There were numerous factors that influenced the duration of the development process of the damage in different stories. These factors include mass distribution, inertia forces and input ground motion.

At times, it is easy to determine the weak locations in the structure by means of determining the internal micro-cracks. This analysis can be used as a tool to determine the early signs of cracking and early warning for a structure that is vulnerable to damage under seismic loading.

Figure 4.21 demonstrates the local strain spectrum of the response monitored from sensor #15. The beam corresponding to sensor #is presented in Figure 4.20b and d. It can be seen from the figure that when the ground motion input peak value is 0.2g and 0.3g, the strain spectrum has obvious frequency peaks. After the local vibration input peak value is greater than or equal to 0.5g, a remarkable frequency shift can be observed. With an increase in the input seismic excitation, it can be seen that the strain spectrum loses its obvious frequency peak. The information given by the strain spectrum indicates that the second-floor beam joint of Figure 4.20d has a damage problem when the ground motion input is increased from 0.5g.

4.6.3   Damage correlation with the dynamic characteristics of RC model

The structural damage response was first observed through variation in the dynamic properties of the structure as reported in Table 4.4. These modal characteristics were determined before each loading state. The reason for high initial frequency in Table 4.4 is due to the stiffness contribution of the shear wall and relatively lower mass of the structure. It can be seen that the first and second order frequencies decreased gradually and the corresponding damping ratio increased simultaneously under progressive seismic loading. This is due to the fact

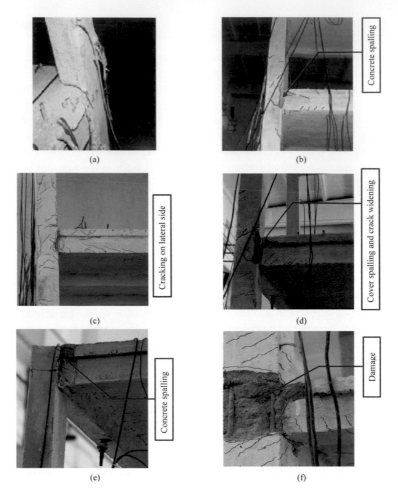

Figure 4.20 Progressive damage near beam-column joint: (a) crack at the beam-column joint of first floor on grid 1-A at PGA = 0.6g; (b) cracks and damage at the beam-column joint of second floor on grid 1-B at PGA = 0.6g; (c) cracks on the lateral side of the flexible edge at PGA = 0.6g; (d) cracks and spalling of concrete cover at beam-column joint of second floor on grid 1-B at PGA = 0.7g; (e) cracks and concrete spalling at beam-column joint of top roof on grid 1-B at PGA = 0.7g; and (f) damage at the beam-column joint at PGA = 0.8g

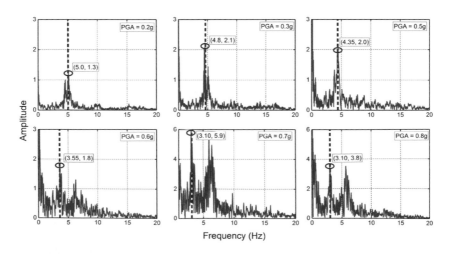

Figure 4.21 Strain spectrum corresponding to the damage near sensor #15

Table 4.4 Dynamic characteristics of RC model

Test	First natural frequency (Hz)	First damping ratio (%)	Second natural frequency (Hz)	Second damping ratio (%)
0.2g	5.55	1.58	24.01	0.67
0.4g	4.73	2.49	21.40	0.96
0.5g	3.59	3.50	20.75	1.52
0.6g	3.10	4.41	18.95	2.15
0.7g	2.94	5.43	18.46	2.77
0.8g	2.61	6.62	17.64	3.73
1.0g	2.10	8.26	16.67	4.75

that the augmentation of the damage in the structure led to the increase in the width of beams and columns; therefore, the increase in the length of the structural members caused significant reduction in the natural frequency of the structure. However, this phenomenon consequently caused significant and rapid rise in the damping ratio of the structure. Summarizing the modal characteristics variation, it can be said that the structure experienced 63% and 30% reduction in the first and second mode natural frequencies in the damaged state respectively, compared with the first and second mode natural frequencies of the structure in the undamaged state. On the other hand, the damping ratio of the structure in the damaged state increased to about 5 times the initial damping ratio of the structure in the undamaged state.

4.7 Summary

The monitored responses obtained through a series of tests prove that FBG sensors have the advantage of significant accuracy, small size and good embedding abilities. It demonstrates its promising failure monitoring abilities and potential for crack detection. The results achieved through this research would be beneficial to validate existing numerical simulation and analytical procedures, particularly for the structures with inherent asymmetry. The main findings of this chapter have been summarized as follows.

4.7.1 Summary of the physical damage in RC model

The three-story asymmetric shear wall structure was exposed to seismic excitations with different PGA levels. The strain in the structure was monitored at different locations using FBG strain sensors. The following experimental observations have been made:

- The torsional rotation in the structure under the influence of seismic excitation affected the response of members located near the edge of the structure, which has led to the conclusion that design of the corner members on the FS needs serious attention.
- The maximum tensile strain in some of the sensors was almost doubled while the compressive strain was increased to nearly 3 times the compressive strain observed in the previous PGA. This phenomenon indicates the presence of a large crack near the sensor region and also indicates that concrete was crushed at this point and eventually confirms the formation of a large crack and crushing of concrete.

- The residual strains of FBG sensors began to increase remarkably when the seismic excitation was increased from 0.5g. This demonstrates the fact that initial micro-cracks appeared inside the structure due to the structure's irreversible elasto-plastic deformation.
- Sudden change in the trend of residual strain was caused by the initial cracks. For most of the residual strain profiles, this sudden change occurred when the excitation was increased from 0.5g to 0.6g which also indicates the initiation of the plastic state of the structure.
- The shear wall region was the strongest zone of the structure and because of the high stiffness of the member, minimal relative strain changes were observed.
- The plastic hinges formation in the beams occurred at PGA of 0.8g while in the columns plastic hinges were formed at PGA of 1.0g.
- In real-life structures, FBG strain sensors have demonstrated great performance in detecting the strain closer to the cracks, thereby providing early warning signs for primary and secondary components of the structure. When the crack is present, the strain close to the crack can be monitored to calculate the maximum crack width in the member. The warning signs for the structures can be established when the maximum crack width is either close to the crack width limit or has exceeded the limit.

4.7.2   Damage simulation in steel models

Through the analysis of the asymmetric steel structures, the seismic response magnifications can be observed under the influence of eccentricity from the monitored strain results. Detailed contour plots presented in this chapter and in Appendix B will enable structural engineers to determine appropriate locations of maintenance where stress concentration is likely to occur. This book provides the quantification of such influences. Concluding the observation, the following points can be drawn:

- In TS structures, the SS is expected to experience higher tensile deformation compared with the FS of the structure. Moreover, higher order floors are expected to get higher damage compared with the lower order floors. However, this is only true when the structure is RI. As soon as the structure becomes an IRI, this behavior slowly fades away and an entirely new pattern of deformation gets developed.
- In TF structures, the FS is expected to experience higher tensile deformations compared with the SS of the structures. Moreover, lower order floors are expected to get higher damage compared with the upper order floors. However, this is only true when the structure is RI. As soon as the structure becomes an IRI, this behavior slowly fades away and an entirely new pattern of deformation gets developed.
- Location-specific eccentricity not only affects the response at the corresponding floor of eccentricity, but it also transmits its influence down to the adjacent floor. Moreover, this influence is more dominant at the SS of the structure. Besides, if the location-specific eccentricity is not at the top roof level, the top roof level is least affected under the influence of vertical mass eccentricity.
- In TF structures, the FS of the structure is expected to experience higher tensile deformations compared with SS of the structures. This behavior is greatly reversed under the influence of location-specific eccentricities. Higher tensile deformations transfer to the SS, whereas the FS remains the least affected. Moreover, this deformation-reversal pattern gets transmitted to the adjacent lower floors as well.

- These contour plots can serve as a structural health monitoring guide for existing asymmetric structures in terms of determining the deformation mode of asymmetric structures of similar types to those considered in this research and to predict the expected location of damage. Moreover, the presented contour plots can provide great help for the design of new structures which are relatively similar to the ones considered in this research in terms of design techniques, such as location of braces in the structural frames, placement of shear walls and collectors for force transfer mechanism.
- These contour plots can help assess the deformation transition from the FS to the SS and from the SS to the FS of the structure at all floor levels.

Chapter 5

Numerical evaluation of complex local behavior

5.1 Introduction

This chapter focuses on the validation of behavior corresponding to the asymmetric distribution of strain within the same column at the FS of C-1 model. Validation of the results from the Finite Element Model was carried out in ABAQUS by comparing asymmetric strain distribution of the FE model with the experimental findings. Moreover, based on the numerical and experimental investigations, the behavior of the local response behavior both at the FS and SS of the structure is discussed and compared. This chapter concludes that corner columns at the FS of such kind of structures require substantial redundancy during the design process as the residual strain behaves entirely different within the same structural member and the torsional vibrations may likely induce excessive shear at the FS-column located at the periphery of the asymmetric structure.

Recent major earthquakes prove the fact that reinforced concrete (RC) structures are prone to collapse under seismic actions (Takewaki *et al.*, 2011). There are several reasons behind the collapse of RC structures under seismic actions, and one such reason is the inaccurate estimation of structural seismic response. This is due to the fact that in asymmetric structures, columns at the periphery of the structure are likely to experience additional shear demands under torsional vibrations. This excessive shear on the vertical column elements is highly dependent on the distance of the flexible column from the C_S, i.e. the higher the distance, the higher the shear demand. Therefore, these corner elements especially at the FS require special attention during the design process. Therefore, a structural collapse can happen even if the design of a structure fully complies with the design guidelines (Osteraas and Krawinkler, 1989; Ger *et al.*, 1993; Minzheng and Yingjie, 2008). The potential for structural failure in asymmetric structures is higher compared with fairly symmetric structures (Oyguc *et al.*, 2018).

5.2 Contribution of this chapter to knowledge

This chapter solely covers the responses from C-1 model and covers two points:

- Experimental and numerical response validation for asymmetric distribution of local response within the same structural member. This issue is considered keeping in view the gradual decline in the local and global resisting capacity of the structure under previous seismic actions, internal cracking and residual deformations.
- Investigation on the behavior of residual strains in the elastic and inelastic states when a structural component is partly under tension and partly under compression.

Covering these issues, experimentally obtained local responses at the FS of the C-1 model were assessed. Moreover, a numerical model was developed and the experimental results were compared with numerical findings. Based on the verified numerical model, local responses at other locations on the FS of the structure were investigated. Finally, a comparison between the abnormal behavior of the local response was compared at both the FS and SS of the structure.

5.3   FBG sensors under consideration

For local response measurement, the installed strain sensors at locations illustrated in Figure 5.1 were used. However, the scope of this book is confined to the varying response within the same structural components, therefore, the structural local response obtained from sensor #1 and sensor #2 (Figs. 5.1a and b) will be discussed here. For deployment of the FBG sensors, several protection measures were taken in order to obtain accurate measurement of the response. These protection measures include the polishing of reinforcement bars using a grinding machine and then careful attachment of the sensor with the reinforcement bars.

5.4   Behavior of local response under progressive seismic excitation

The neutral axis is the term in material mechanics where the strain monitored by the strain sensor is defined corresponding to a "neutral axis" under a single ground motion. At the end of the vibration if the strain is consistent with the initial strain, that is known as unchanged neutral axis. If the strain at the end of the vibration is higher than the initial strain or below

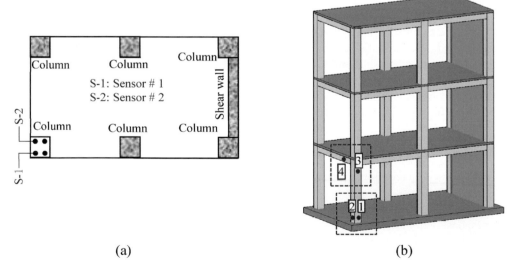

(a)   (b)

Figure 5.1 Selected column at the flexible edge of the structure for strain monitoring: (a) location of FBG sensors under consideration and (b) location of FBG sensors at other locations for validation of experimental results with numerical results

the initial strain, that is defined as neutral axis up and neutral axis down. Also, if the neutral axis has been declining under the action of several earthquakes, it indicates the presence of cracks. The location of these cracks would be in the vicinity of the sensors because the emergence of cracks would cause the decline of the neutral axis due to the release of tensile strain acting on the sensor. The frequent movement of the neutral axis under the action of ground motion indicates the crack near the sensor region as well. The tensile strain in that crack part is termed as residual strain. Considering the experimental results, the neutral axis term was considered in assessing the structural local response. The two described situations where residual strain could form either in the negative or positive region when the structure changes its states from elastic to plastic state forming either internal micro-cracks or transferring into plastic state were evaluated from Figures 5.2–5.4.

The described complex situation occurs when the reinforced elastically deformed structure restores a portion of the plastic deformation under the action of ground motion. This analysis has been made only for the sensors installed in the corner column at the FS of the structure. Also, in strain history profiles, the positive strain corresponds to the tensile strain while the negative strain corresponds to the compressive strain.

5.4.1 Behavior of local response in the elastic state

The observations made in this section from strain history profiles do not consider the initial strain caused by the casting and pouring of concrete. The FBG sensors labeled "sensor #1" and "sensor #2" with their location being in the same column but with different reinforcement bars have been selected to analyze the influence of asymmetry on the FS of the structure. Figure 5.2 shows the strain time history responses of the selected sensors for the input ground motions of 0.2g and 0.3g respectively, in the elastic state.

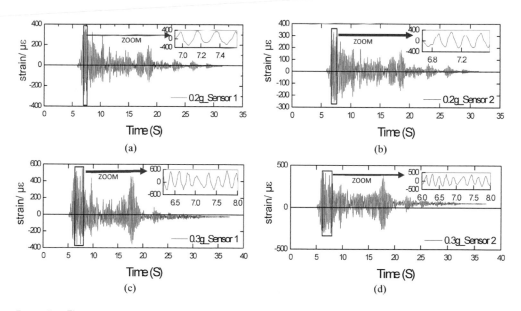

Figure 5.2 Elastic strain response at flexible edge under progressive loading: (a) PGA = 0.2g for sensor #1; (b) PGA = 0.2g for sensor #2; (c) PGA = 0.3g for sensor #1; and (d) 0.3g for sensor #2

It can be seen that the strains of both sensor #1 and sensor #2 increase gradually with the increase in the input ground motion, when the ground acceleration was increased from 0.2g to 0.3g. For the first input excitation of 0.2g, it can be noticed that both the sensors reported unequal tensile and compressive strains despite of the fact that both the sensors were installed within the same column. Sensor #1 has around 25% higher compressive and tensile strains than sensor #2. This is because of the contribution of asymmetry and consequences of irregularity on the FS of the structure. It can also be noted that sensor #1 achieved some part of residual strain and the neutral axis moved up when the seismic excitation was 0.2g. When the input acceleration was further increased from 0.2g to 0.3g, the scenario of neutral axis up and down clearly formed. At this PGA level, the neutral axis moved down for sensor #1 and moved up for sensor #2, giving rise to a completely different behavior of the stored strains within the same structural member. It can further be noticed from Figure 5.2 that tensile strain is reasonably higher than compressive strain for sensor #1. However, both tensile and compressive strains are fairly equal for sensor #2. Furthermore, the compressive and tensile strain of sensor #1 is around 25% higher than the compressive and tensile strain of sensor #2. As a matter of fact, the structure is still intact and is in the elastic state with no apparent and internal damage. This led us to the conclusion that asymmetry of the structure can lead to the non-uniform responses of the same structural component even within the elastic state.

5.4.2   Behavior of local response at internal micro-cracking state

When the frame-shear wall structure was exposed to seismic excitation of 0.5g, some very interesting phenomena were noticed. No apparent cracks were noticed in the structure at this stage; however, the structure's critical state started to begin from this point. Looking at the local response of the structure at this point (Fig. 5.3) and at the variation in the dynamic properties of the structure (Fig. 5.5), it can be said that internal micro-cracks started to form at this loading stage. Further, in terms of local seismic response, it can be seen that negligible progressive increase in the compressive strain of sensor #1 is observed while the tensile strain almost doubled for the same sensor. On the other hand, tensile and compressive strains increased with the same proportion for sensor #2 and equal tensile and compressive strain was recorded for this specific sensor. Comparing the response obtained through both the sensors, the tensile strain for sensor #1 is still higher than the tensile strain in sensor #2 while the compressive strain in sensor #1 turns out to be lower than the compressive strain of sensor #2 at this PGA.

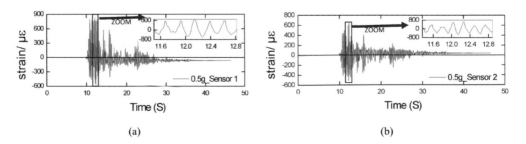

(a)   (b)

Figure 5.3 Strain response at flexible edge in the internal micro-cracking state: (a) PGA = 0.5g for sensor #1 and (b) PGA = 0.5g for sensor #2

5.4.3   Behavior of local response in the inelastic state

When the seismic input acceleration was increased from 0.5g to 0.6g, the structure entered into the plastic state and started to form narrow cracks near the beam and column ends. This accelerated the structural non-uniform response within the same structural component. Strain history responses for the plastic state of the structure are illustrated in Figure 5.4. It

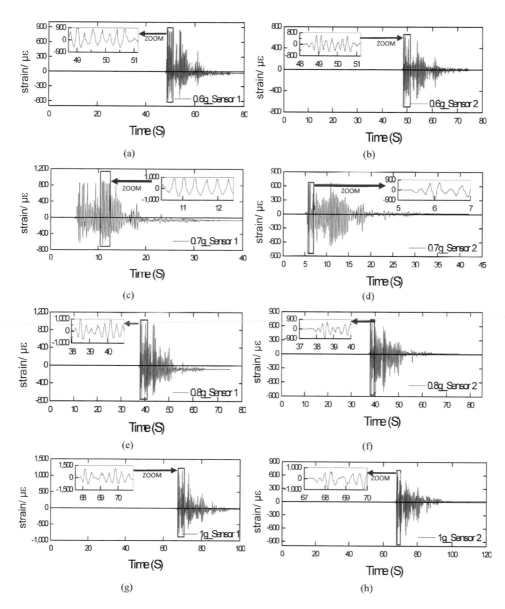

Figure 5.4 Plastic strain response at flexible edge under progressive loading: (a) PGA = 0.6g for sensor #1; (b) PGA = 0.6g for sensor #2; (c) PGA = 0.7g for sensor #1; (d) PGA = 0.7g for sensor #2; (e) PGA = 0.8g for sensor #1; (f) PGA = 0.8g for sensor #2; (g) PGA = 1.0g for sensor #1; and (h) PGA = 1.0g for sensor #2

can be seen that the structure's local response at PGA = 0.6 has given very unusual behavior in terms of neutral axis position. This is due to the fact that cracks help in releasing some part of the strain which eventually influences the neutral axis scenario. Looking at the strain responses of sensor #1 and sensor #2, it can be seen that there is almost negligible rise in the compressive and tensile strains for both the sensors. Because of the formation of cracks in the structure, some part of the strains has now started to release through the cracks. This fact is noticeable for all the PGAs from now onwards. At PGA of 0.6g, both the sensors have fairly equal compressive strains. However, the tensile strain of sensor #1 is around 25% higher than the tensile strain of sensor #2. When the input acceleration was further increased from 0.6g to 0.7g, a small increase in the compressive and tensile strain occurred for both the sensors. However, the compressive strain of sensor #2 began to increase from the tensile strain while the compressive and tensile strains of sensor #1 reported fairly the same trend in the response. It can be seen that both the sensors were installed within the same structural member, but the structural member seems partly under tension and partly under compression. For input excitations of 0.8g and 1.0g, the local response followed the same trend as it did for PGA of 0.7g with a little difference that compressive and tensile strains started to decline at PGA of 1.0g. Since the structure at this stage is completely in the plastic state, this decline also reports the presence of cracks near the sensors location.

Overall, it can be said that maximum tensile strain of sensor #1 was found to be greater than the maximum tensile strain of sensor #2, while the maximum compressive strain for sensor #1 was greater than the maximum compressive strain for sensor #2 until the PGA was 0.6g. Further increase in the PGA reported fairly equal strains for both the sensors. It can also be concluded that different seismic local responses of different reinforcement bars but within the same column reveals the need for careful attention towards the corner columns during the design process. However, in fairly symmetric structures, this phenomenon is very unlikely to occur and the reason is explained in Figure 5.14 in Section 5.6, where it can be seen that the corner column at this edge experienced fairly uniform behavior within the same structural member. The maximum strain at the same height of the same column is consistent.

5.5   Damage in terms of residual strain and variation in the dynamic properties

5.5.1   Correlation with varying dynamic properties

Figure 5.5 illustrates the variation in the dynamic properties of the structure investigated after each loading state. Structural responses using white noise excitation were obtained, which were then transformed into frequency response functions to calculate structural dynamic properties. Since the structure was exposed to progressive seismic excitations from 0.2g to 1.0g with an increment of 0.1g, the structure experienced several deformation states which eventually influenced the dynamic properties of the structure. It can be seen that the structure experienced a gradual reduction in the frequency of the structure in first and second mode under progressive seismic load with highest frequency reduction of 65% in the first mode after the final loading state. However, it is evident that the frequency reduction is only higher in the first mode as compared to the second mode where the structure experienced relatively lower frequency reduction of 35% after the final loading state. This may be due to the fact that the structure has a low damping ratio in the higher modes. Conversely, looking at the damping ratio of the structure, it can be seen that the damping ratio increased gradually

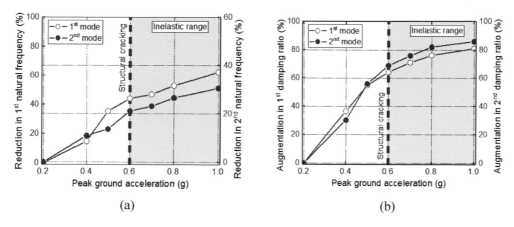

Figure 5.5 Variation in the dynamic characteristics of the structure: (a) natural frequency and (b) damping ratio

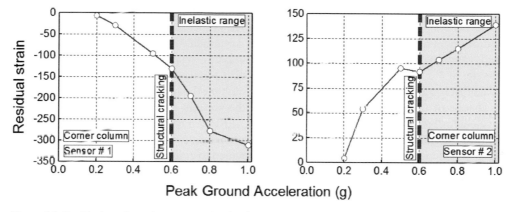

Figure 5.6 Residual strain under progressive loading corresponding to sensor #1 and sensor #2

with decrease in the frequency of the structure. However, the damping ratio augmentation is relatively equal in first and second mode having an approximate augmentation of 80% and 85% in the two modes. Numerous other researchers have found similar observations which confirms the facts stated in this section (Xiao *et al.*, 2015; Cao *et al.*, 2009; Xu *et al.*, 2018; Kim *et al.*, 2012; Li *et al.*, 2016).

5.5.2   Damage at FS in terms of residual strain

In Figure 5.6, the residual strains obtained through the sensors installed in the selected column have been graphically presented. Taking for example sensor #1 and sensor #2 for the evaluation of residual strain, it can be seen that residual strain in sensor #1 was negatively increased under progressive earthquake loading while residual strain in sensor #2 was increased positively under progressive seismic excitations. This gives rise to the fact

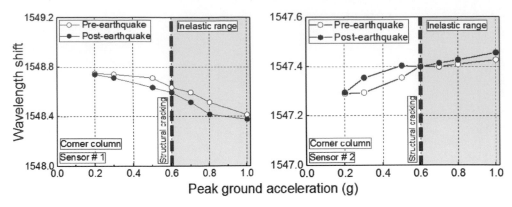

Figure 5.7 Pre and post-earthquake wavelength shift in sensor #1 and sensor #2

that the trend of residual strain is completely opposite in both the sensors, but interestingly, both the sensors were installed in the same column. However, the sensors were attached to different reinforcement bars. This phenomenon of a completely opposite trend of residual strains obtained from the sensors that were installed in the same column but with different reinforcement bars was due to the involvement of asymmetry in the structure and hence, the trends confirm the influence of asymmetry on structural response due to lateral-torsional behavior at the FS of the structure.

The FBG strain sensors provided the residual strain on the reinforcing bars by comparing the change of the wavelengths before and after the seismic excitation. The change in the wavelengths before and after the seismic action is illustrated in Figure 5.7. Since this experiment was an indoor experiment, the influence of temperature was eliminated.

The wavelengths of FBG sensors have been provided to consider the initial strain (shrinkage strain). The provided strain profiles have been monitored without considering initial strain due to curing and pouring. For more reliable results and better comparisons, these wavelengths can be used in correspondence with the temperature to obtain the initial strain of the overall structure before the test using Equation 4.5.

For sensor #2, considering the pouring temperature as 25 degree Celsius, test temperature as 5 degree Celsius, center wavelength of the FBG sensor before pouring as 1547.797nm, and before the test as 1547.289nm, temperature sensitivity co-efficient as 0.027nm/°C and strain sensitivity co-efficient as 1.2×10^{-3} nm/μɛ, the initial strain turns out to be 26.6μɛ. Similarly, considering the same procedure for sensor #1, the initial strain turns out to be 46.9μɛ.

By considering the influence of that calculated initial strain, more reliable analytical data can be demonstrated for the analysis of either yielding of steel bars or ultimate tensile strain of steel bars.

5.6   Finite element modeling of RC model

For further evaluation of the experimental response of RC model, FE model was aimed to be developed in finite element software ABAQUS. Damage plasticity models are widely used in material modeling of concrete. In damage mechanics, concrete fractures are simulated by stiffness degradation due to the formation of micro-cracks. Numerous concrete damage plasticity models have been developed by researchers in the past. However, the nature of concrete is composite; its complex behavior specifically under cyclic loadings cannot be reflected in constitutive theory of material as like plasticity theory and damage mechanics theory (Lee and Fenves, 1998). The material properties for structural elements were used in accordance with the experimental data in the form of the engineering stress (σ_E) and strain (ε_E). In finite element analysis, the term true stress (σ_T) and the true plastic strain (ε_T^{pl}) (ABAQUS, 2003) are used to define the material properties. Engineering stress and logarithmic strain relationship can be used to obtain true stress and true plastic strains as expressed in Equations 5.1 and 5.2:

$$\sigma_T = \sigma_E \left(1 + \varepsilon_E\right) \tag{5.1}$$

$$\varepsilon_T^{pl} = \ln\left(1 + \varepsilon_E\right) - \frac{\sigma_T}{E} \tag{5.2}$$

where E is the elastic modulus. Concrete is a complex non-homogeneous material as compared to other structural materials such as steel. The behavior of this complex material can be assigned in ABAQUS using either concrete smeared cracking model, Drucker Prager model or concrete damaged plasticity model. The concrete damaged plasticity model has been adopted here in this chapter. Damage plasticity model is useful in predicting the behavior of the structure under seismic ground motions. Concrete damage plasticity is defined using stress-strain relations and damage parameter for tension and compression (Fig. 5.8a and b). Nonlinear dynamic analysis has been considered in this chapter to evaluate the dynamic behavior of structures. Four major material parameters are used in the concrete damaged plasticity model: the dilation angle (ψ), eccentricity (e), ratio of the second stress invariant on the tensile meridian to that on the compressive meridian (K_c) and ratio of the compressive strength under biaxial loading to uniaxial compressive strength (f_{b0}/f_{c0}). Poisson's ratio (v_c) was considered as 0.2 and default values of 30°, 0.1, 1.16 and 0.67 for ψ, e, K_c and f_{b0}/f_{c0} respectively, were used for modeling the asymmetric reinforced concrete structure. The elasticity modulus of concrete (E_c) was calculated from $4700\sqrt{fc'}$ as recommended by ACI-318 (2008).

The compressive stress-strain relationship of concrete was defined in terms of stresses (σ_C), inelastic strain (ε_C^{in}) corresponding to the stress value and compressive damage parameters (d_c) with inelastic strain values. Equation 5.3 can be used to convert total strain values to inelastic strains:

$$\varepsilon_{in}^c = \varepsilon_c + \varepsilon_{el}^c \tag{5.3}$$

where ε_c is the total compressive strain, ε_{in}^c is the inelastic compressive strain of concrete and ε_{el}^c is the elastic compressive strain of the undamaged concrete and equals the ratio of elastic compressive stress to the initial elastic modulus of concrete. In this work, the compressive stress-strain curve was utilized from Hsu and Hsu (1994). The compressive stress-strain curve was obtained from an experimentally validated numerical method by Hsu and Hsu (1994). The model is useful in developing stress-strain relationship under uniaxial compression until $0.3\,f_c'$ in decreasing portion only by using maximum compressive strength (σ_{cu}). Figure 5.8a defines the ultimate compressive stress (f_c'), strain at f_c' (ε_0) and corresponding strain to stress at $0.3f_c'$ in downward decreasing portion (ε_d). A stress-strain relationship compliant to the Hooke's law is assumed up to $0.5f_c'$ in ascending portion. Hsu and Hsu's numerical model is utilized only to obtain compressive stress (σ_c) between $0.5f_c'$ and the $0.3f_c'$ in the descending portion using Equation 5.4:

$$\delta_c = \frac{f_c' \, \beta'(\varepsilon_c / \varepsilon_o)}{\beta - 1 + (\varepsilon_c / \varepsilon_o)^\beta} \tag{5.4}$$

The strain ε_d is iteratively calculated using Equation 5.4. The parameter β' depends upon the shape of the stress-strain diagram and is derived from Equation 5.5. The strain ε_o and modulus of elasticity ε_0 at peak stress are given by Equations 5.6 and 5.7:

$$\beta' = \frac{1}{1 - \left[f_c' / (\varepsilon_o E_o)\right]} \tag{5.5}$$

$$\varepsilon_o = 8.9 \times 10^{-5} f_c' + 2.114 \times 10^{-3} \tag{5.6}$$

$$E_o = 1.234 \times 10^{-2} f_c' + 3.28312 \times 10^3 \tag{5.7}$$

For tensile behavior of reinforced concrete, a stress-strain relationship after failure for concrete under tension (Fig. 5.8b) was developed to consider reinforcement interaction with concrete, strain softening and tension stiffening. Modulus of elasticity (E), cracking strain (ε_{cr}) and true stress (σ_T) were used for the considered grade of concrete to develop this model where cracking strain (ε_{cr}) can be calculated from the total strain using Equation 5.8:

$$\varepsilon_{ck.}^{ten.} = \varepsilon_{tot.}^{ten.} + \varepsilon_{el.}^{ten.} \tag{5.8}$$

where $\varepsilon_{ck.}^{ten.}$ is cracking tensile strain, $\varepsilon_{tot.}^{ten.}$ is the total tensile strain and $\varepsilon_{el.}^{ten.}$ is the elastic tensile strain of the undamaged concrete material, which equals the ratio of elastic tensile stress to the initial modulus of elasticity of concrete. There are different forms of tension stiffening models presented in the literature as reviewed in the paper by Nayal and Rasheed (2006).

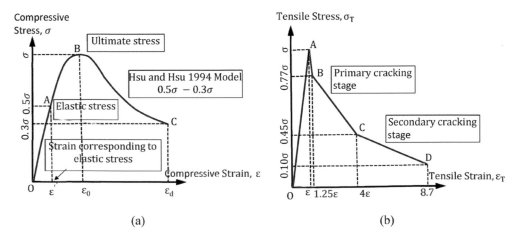

Figure 5.8 Damage plasticity model: (a) compressive damage model and (b) tension stiffening model

The model developed by Nayal and Rasheed (2006) is selected for the present study, as it is applicable to RC structures with only minor changes to avoid convergence (Wahalathantri *et al.*, 2011). The presented tension stiffening model is originally based on the stress-strain relationship established by Gilbert and Warner (1978). Two decreasing areas of the tensile behavior graph have addressed the response due to primary cracking phenomenon and secondary cracking phenomenon (Gilbert and Warner, 1978). The modified tension stiffening model used in this chapter is shown in Figure 5.8b. There are four parts in the stated model: the un-cracked stage (OA), the primary cracking stage (AB and BC) and the secondary cracking stage (CD). Using this equation, convergence during FE analysis is raised. The proposed model by Nayal and Rasheed is utilized in this chapter to overcome the issues of convergence as this model provided a best fit to the convergence issue. At critical tensile strain ε_{cr}, there is a sudden drop from maximum tensile stress σ to 0.8σ as used by Nayal and Rasheed (2006) and Gilbert and Warner (1978) and is slanted from (ε_{cr}, σ) to ($1.25\varepsilon_{cr}$, 0.77σ) to avoid run-time errors in the ABAQUS damage plasticity model. The path of stress-strain curve used in this chapter is exactly similar to the described model in primary and secondary cracking region but in order to avoid ABAQUS runtime errors the curve is stopped at ($8.7\varepsilon_{cr}$, 0.10σ).

5.6.1 Development of the FE model

The simulation is based on the 3D modeling of the reinforced concrete structure used for the experiment as shown in Figures 5.9–5.11. The eight-node iso-parametric solid and truss elements having two integration points in each direction are utilized to complete the numerical modeling. The structure has been modeled with a total number of nodes of 46,747 and

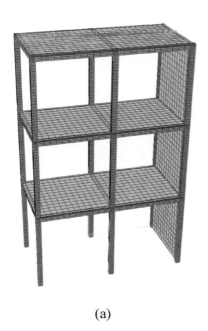

(a)   (b)

Figure 5.9 Reinforcement and meshing details in FE model: (a) meshed model and (b) reinforcement in the FE model

with total number of elements of 38,868. The concrete has been modeled using solid element C3D8R, i.e three-dimensional, eight-nodded brick elements with reduced integration technique while horizontal and vertical reinforcement bars have been modeled as T3D2, i.e. two-nodded, three-dimensional truss elements. The truss elements have been meshed to form 22,992 linear line elements of type T3D2, and the solid elements have been meshed to form 15,876 linear hexahedral elements of type C3D8R. A perfect bonding condition has been assumed between the concrete and the reinforcement bars. The stiffness proportional damping factor is computed such that 5% damping is being captured for the fundamental vibration period. At the first step of loading, the structure is subjected to gravity load, and then the El Centro ground motion is applied to the structure to capture the seismic response under dynamic load.

The base of the columns was restrained in all possible directions except the direction of input acceleration. The top of the structure was made free to move in any direction. During the experiment, the asymmetric structure experienced two kinds of loads: (1) gravity loads in the downward direction and (2) ground motion excitation along the transverse direction of the asymmetric structure. Therefore, in the FE modeling, these two loading conditions were considered to capture the actual conditions experienced by the structure during the experiment. The loading in the structure was considered in two steps. Initially, the reference point was constrained along all the directions except the loading direction of the ground motion.

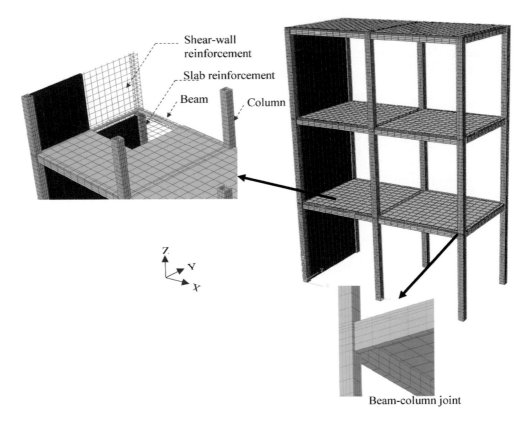

Figure 5.10 Finite Element Model of the asymmetric structure

(a)   (b)

Figure 5.11 FE modeling for local response evaluation: (a) reinforcement embedment in the concrete sections and (b) surfaces

Then, the gravity loads were assigned to the structure under a boundary condition of seismic excitation.

As shown in Figure 5.11a, the reinforcement bars have been embedded into the concrete. The reinforcement is considered as embedded region while the concrete is considered as the host region. Both absolute and fractional tolerance has been considered with absolute fractional tolerance being zero and fractional exterior tolerance being 0.05. Tie contact between the structural components was generated using a surface to surface discretization method. In Figure 5.11b, it can be seen that a total of 126 tie surfaces were generated to consider the contact between the structural components. Ties in this FE model comprise of the ties between column-to-shear, beam-to-shear wall, column-to-slab, beam-to-slab and column-to-beam. While defining the contact between the structural components, the master and slave surfaces are required to be carefully defined for contact formulation. The master surfaces are defined as surfaces belonging to the body of the stronger material, or bodies of a finer mesh.

5.6.2   Numerical response validation

The local response of the structure was investigated in finite element model. The responses were first compared with experimentally obtained strain responses at strategic locations. The local responses have been compared for a loading state of PGA = 0.2g in the elastic state and for a PGA of 1.0g in the plastic state. The numerical responses have been found to be in moderate agreement with experimental results. Therefore, these numerical models have been considered valid to carry out further investigation. In Figure 5.12, the locations of the numerical strain responses have been demonstrated. The strains obtained from the described locations have been compared with experimentally obtained strains from FBG sensors.

Figure 5.13 describes the comparison between the experimental and numerical results in the elastic state. A non-linear FE model of the frame-shear wall structure was simulated in ABAQUS. However, the numerical model was first examined in the elastic state as at PGA

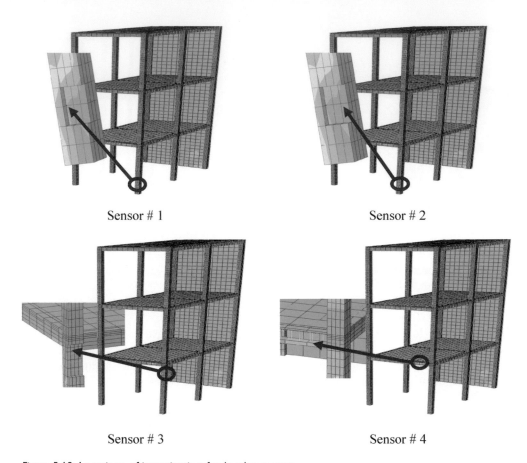

Sensor # 1   Sensor # 2

Sensor # 3   Sensor # 4

Figure 5.12 Locations of investigation for local response

of 0.2g where structure remained in the elastic state. Therefore, an elastic analysis in the FE model was carried out at PGA of 0.2g. Figure 5.13a demonstrates that the numerical results in the elastic state are in agreement with the experimental results to a certain extent.

For the plastic analysis, the non-linearity of the structure was established in the form of concrete and reinforcement's strain-strain curves as explained earlier. The achieved numerical results in Figure 5.13 illustrate the comparison between the experimental and numerical results when the input ground motion was 1.0g. It can be seen that the numerical results are in good agreement with the experimental results. Although the numerical results match reasonably well with the experimental results, there is some amount of errors at some peak shifts. Illustrations of Figure 5.13 report that numerical results are slightly higher than the experimental at a few locations especially when the input ground motion is 1.0g. Furthermore, Figure 5.13 also illustrates that numerical results are slightly lower than the experimental results at PGA = 0.2g. Due to the highly dynamic nature of the structure in the inelastic state, errors in the numerical results were expected beforehand. In addition, the material characteristics of the actual constructed structure may have differed a bit with the material strengths due to poor construction practice. Other reasons are explained in the next paragraph.

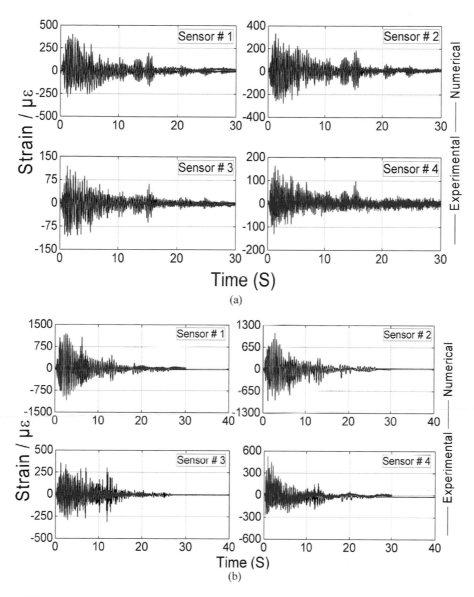

Figure 5.13 Comparison of the experimental and numerical strain response at (a) PGA = 0.2g and (b) PGA = 1.0g

The established FE model demonstrated moderately agreeable results for validation purposes. However, it is evident that the results did not perfectly match with the experimental findings mainly because of the assumptions considered in the modeling process. Given that the primary goal of FE modeling is the conceptual comparison of the local responses with experimental results in the elastic and inelastic states, the established FE model was found appropriate for approximate evaluation of the nonlinear seismic response. One of the main limitations of the modeling process was the model updating based on the damage that occurred in the precious loading state. The reason why model updating was not considered in this research is subject to the involvement of numerous uncertain parameters affecting the seismic response such as non-uniform yielding, varying center of stiffness due to damaged structural components, varying center of mass and varying modal properties of the structure. Since the accuracy of the updated FE model requires higher computational intelligence and less uncertain parameters, the updated FE model established, merely on the basis of engineering judgment, could have led to a biased structural response in this particular case. Therefore, the numerical investigation in this research did not utilize the updated FE model because of the complexity of the scenario and numerous uncertain parameters influencing the seismic response. This chapter reflects non-uniform damage-accumulation caused by the progressive seismic excitation, which is highly complex to be truly implemented in the numerical model. Besides, in the actual structure, there always exist some weak parts, when the smaller forces cause internal cracking. The numerical response considers the larger force of the crack and not the actual existence of the weak parts. This yields in the difference between the actual situation and theoretical assumptions in the cracking time and place. In addition, the sliding between the model beam and the vibration table during the experiment also brings an error to the experiment, which is not reflected in the numerical response. The deformation of the beam-column joints has not been considered in the numerical results, which eventually had an impact on the results. On contrary, the achieved numerical results demonstrated a moderate agreement in the numerical and experimental results and therefore, in the opinion of the author, such numerical models can still be used for approximate evaluation of the nonlinear seismic response as has been done by several other researchers (Li *et al.*, 2006; Lu *et al.*, 2016; Lee and Woo, 2002).

5.7   Comparison of numerical response at FS and SS

To confirm the observations made during the experimental investigation of the structure, local responses in the FE model were observed at the FS of the structure. The local responses from the SS were also obtained to compare the behavior of local response at the SS and FS altogether. Figure 5.14 illustrates the local response in the corner column of the structure at the FS and SS. The responses were recorded in a similar manner as they were recorded during the experiment within the same structural component. Two different reinforcements running in the same direction within the same structural components were kept under focus at the base, first-floor and second-floor level. The locations for the strain responses were observed at the similar points so that both the locations may have an equal comparison of results. Similar criteria were fixed with similar locations at the SS of the structure. From the achieved results, it can be said that the influence of the local response is abnormal at the FS only compared with the local response at the SS of the structure. Considering the local response at the FS of the structure, Figure 5.14 concludes that non-uniform local seismic response within the same structural component maintained its abnormality at all floor levels. However, looking for the same fact at the SS of the structure, it can be seen that local response within the same structural component is uniform with a negligible difference in the responses. Therefore, it can be concluded from these responses that under the influence of irregularity, the FS tends to produce varying local response within the same structural components. The response at the SS, however, remains the same with negligible variation. Figure 5.14 also leads to an important fact that despite

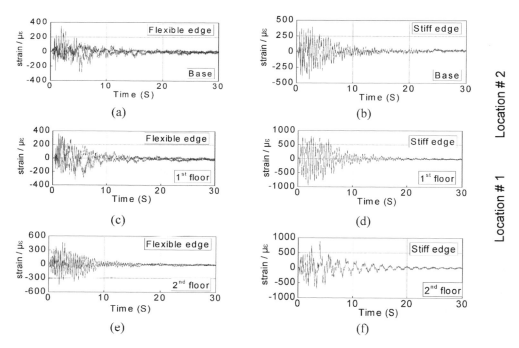

Figure 5.14 Local response in the corner column at the FS and SS: (a) local response corresponding to the base level at the flexible edge; (b) local response corresponding to the base level at the stiff edge; (c) local response corresponding to the first-floor level at the flexible edge; (d) local response corresponding to the first-floor level at the stiff edge; (e) local response corresponding to the second-floor level at the flexible edge; and (f) local response corresponding to the second-floor level at the stiff edge (location #1 corresponds to the first rebar within the corner column; location #2 corresponds to the second rebar within the corner column)

the uniform local seismic response in the corner column of the SS, the structural element is likely to experience higher axial deformations compared with the corner column of the FS as has been experienced in this case.

5.8 Summary

This chapter discusses the asymmetric distribution of strain response in the corner element at the FS of the C-1 model under elastic as well as highly inelastic states. Validated responses from the FE model are used to establish the following observations:

- Corner columns at the FS of the asymmetric structures are likely to remain partly under tension and partly under compression in terms of stored deformations. Conversely, compared with the FS, corner columns at the SS of the structure may not experience this varying trend in the local response. However, the corner columns at the SS of the structure may likely experience higher axial deformations as compared to the FS of the structure.

- Different compressive and tensile strains of different reinforcement bars but in the same column confirm the abnormal behavior of the local seismic response at the FS of the structure. The main reason why the SS of the asymmetric structure didn't experience this phenomenon is the ability of the structural components of the SS to remain in the same state and maintain a uniform stress variation.
- The phenomenon of a completely opposite trend of the residual strains obtained from the FBG sensors that were installed in the same column but with different reinforcement bars was due to the involvement of asymmetry in the structure and hence, the trends confirm the influence of asymmetry on the structural local response.
- This study can be helpful in understanding the local seismic behavior and further development of a practical engineering problem on local collapse mechanism in the vertical corner elements (Lu *et al.*, 2016; Pearson and Delatte, 2005).

For validation of the numerical results with experimental findings, several simplifications in the FE model were considered to predict the local seismic response. These simplifications caused the error at some peak shifts in the simulated results. For example, for simplification purpose, the geometric nonlinearities, buckling behavior and deformation in the beam-column joint were not considered in the FE modeling. Moreover, the damage accumulation of the structure caused by the input of the shaking table is very complex to be truly reflected in the numerical model and this eventually influenced the seismic response of the simulated model. Besides, in the actual structure, there always exist some weak parts, when the smaller forces cause internal cracking. The numerical response considers the larger force of the crack and not the actual existence of the weak parts. This yields in the difference between the actual situation and theoretical assumptions of the finite element modeling.

In a nutshell, the numerical and experimental results are in reasonably good agreement to a certain extent and this indicates that the numerical model can be used for comparison purposes and for further development of the studies on behavior of the local seismic response.

Chapter 6

Global seismic behavior of asymmetric building structures

6.1 Introduction

This chapter provides a detailed investigation on the global behavior of asymmetric structures. Since model C-1 was excited until the plastic hinges formed near the beam-column joints, global responses of C-1 model have been explained in regards to its correlation with the damaged state of the structure. Moreover, a detailed investigation on the influence of seismic response under torsional vibrations has been discussed for steel models.

6.2 Contribution of this chapter to knowledge

As discussed in Chapter 2, previous research on the investigation of seismic response in asymmetric structures is either limited to single-story structures or it has been based on an overly simplified single-story model. It has also been discussed in Chapter 2 of this book that previously obtained findings are not representative of the actual multi-story structures. In this regards, this chapter provides a detailed investigation on the global behavior of vertical as well as planar asymmetric structures. Based on the elastic and inelastic responses of C-1 model, it has been observed that:

- Irregularities in a structure can lead to significant augmentation of the dynamic response due to induced torsional vibrations.
- The damage in the structure influences the dynamic properties of the structure which eventually tend to influence the structural global response.
- Structural acceleration response is more sensitive to seismic excitations and appears to have more influence on local oscillations compared with structural displacement response.

Based on the global responses of steel models, it has been observed that:

- In TF structures, strength irregularity is a more sensitive parameter under torsional vibrations compared with stiffness or mass irregularity.
- In TS structures, the highest influence of the torsional vibrations is at the FS of the structure whereas in TF structures, the highest influence of the torsional vibrations is at the SS of the structure.

Detailed observations are presented at the end of this chapter. It is interesting to note that each seismic demand parameter has a strong dependency on three parameters: (1) ground motion characteristics, (2) excited mode of the structure and (3) type and location of inherent eccentricity of the structure.

6.3   Dynamic acceleration response

6.3.1   Elastic and inelastic acceleration response of RC model

The achieved maximum acceleration value was arranged on the base model of the accelerometer as the reference standard and then the structure of each model was compared to a corresponding accelerometer direction acceleration measured at the peak value of the base peak observed in the same conditions. Figure 6.1 shows the acceleration response of the considered frame-shear wall asymmetric structure. The structure was observed to be completely in the elastic state until the PGA was 0.3g. Exceeding the PGA from 0.3g to 0.5g, the structure started forming micro-cracks. However, the structure didn't reach the inelastic state as yet. The acceleration responses have been presented here both for the elastic and inelastic state of the structure. The structure under observation did not form any visible cracks until the PGA was 0.6g; therefore, it can be seen that within the elastic state the structure's acceleration response was quite different for PGAs of 0.4g and 0.5g. During this

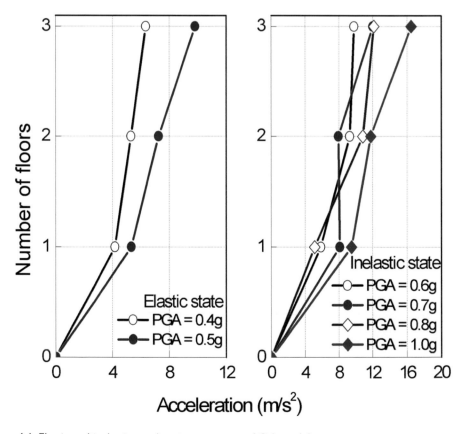

Figure 6.1 Elastic and inelastic acceleration response of C-1 model

state, there was a consistency in the acceleration response with the highest response being on the top floor while the lowest response was reported on the first floor. After 0.5g, when ground input acceleration was further increased, the structure started forming visible cracks and the formation of these cracks significantly influenced the structural response. Since asymmetry of the structure has contributed to the non-uniform yielding of the structure, the structure's response can be seen in Figure 6.1 with fairly equal structural acceleration of 9.1 m/s^2 on the second and third floor. This is because at this point, significant degradation in the structural dynamic properties was noticed. This degradation in the dynamic properties eventually influenced the structural response and thus, after the structure cracking occurred, sudden change in the structural responses and the height-wise variation in the response started to occur. The non-uniform yielding in the structure further increased when the input ground acceleration was augmented from 0.6g to 0.7g. However, the structural acceleration response formed fairly similar behavior as it did for the ground input acceleration of 0.6g. This ensured the fact that the structure is now in the plastic state. For a PGA of 0.7g, the lowest acceleration response was recorded on the first level floor while the second and third floors appeared to have fairly equal acceleration responses. When the ground input motion was further increased to 0.8g, the non-uniform yielding in the structure caused significant reduction in the story stiffness which is evident from the structural acceleration responses. At a PGA of 0.8g, the structural response deviated from its response trend for previous PGAs. This time the maximum structural acceleration of 12.5m/s^2 was recorded on the top floor and lowest structural acceleration was recorded on the second floor instead of first floor. The structural acceleration response for the second floor at 0.8g was even lower than the responses obtained from PGAs of 0.6g and 0.7g. This is due to the fact that at this stage, the structure's dynamic characteristics were greatly influenced due to the development of wide cracks and plastic hinges in the beams. Finally, the structural acceleration response for a PGA of 1g followed the same pattern as it did for a PGA of 0.8g. However, this time the non-uniform yielding caused the columns to form plastic hinges. It was found that structural maximum acceleration response occurred at the top floor and minimum response at the bottom floor.

It can be concluded from the experimental floor acceleration response that under progressive increase in the PGA, the structural cracking increased, the stiffness of the structure decreased and the seismic response increased with abnormal patterns due the development of cracks and non-uniform yielding.

6.3.2   Acceleration response of bi-eccentric S-1 model

As discussed earlier, the major effect of the simulated earthquake response on the TU steel models is to cause torsion under varying eccentricities. This chapter is mainly concerned with assessing the response in terms of global behavior considerations and the influence of torsional moments present in the TU-TF and TU-TS structures. The test results also reflect the effective pattern recognition system of the global demands without appreciable damage in the structure under varying eccentricities.

In the case of the bi-eccentric S-1 model in Figure 6.2, it can be seen that the acceleration demands at the two edges have shown similar response trends under near-field and far-field seismic excitation. Unlike angular drifts, acceleration demands have been found to be higher at the FS compared with the SS of the structure for all cases of asymmetries. In particular, from case 4 and case 7, it can be seen that the top floor eccentricity has the highest influence

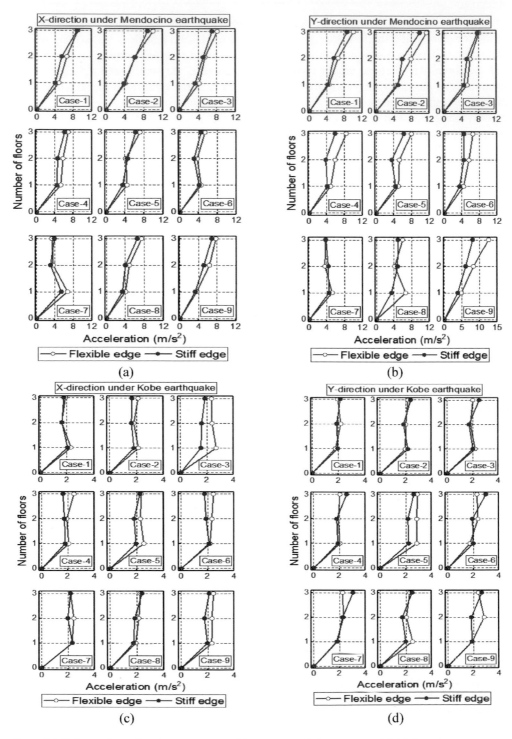

Figure 6.2 Enhanced acceleration response of bi-eccentric S-1 structure: (a) X-direction response under Mendocino earthquake, (b) Y-direction response under Mendocino earthquake, (c) X-direction response under Kobe earthquake and (d) Y-direction response under Kobe earthquake

on maximum acceleration demands at the first-floor level for both near-field and far-field seismic excitation. On the other hand, lower floor eccentricities (case 2 and case 7) have the highest influence on the top floor's acceleration demands under both near-field and far-field seismic excitation. However, this influence is more sensitive at the FS of the structure compared with the SS of the structure. Differentiating between the seismic demands of RI (case 1 and case 5) and IRI scenarios (all cases except case 1 and case 5), it can be seen that response transition from the FS towards the SS is more widespread at the top floor level in RI structures compared with IRI structures. This in turn gives rise to the fact that the top floor's response is more enhanced if all the floors carry approximately equal amount of planar eccentricity along the height of the structure. Furthermore, it can be seen that when the intensity of eccentricity (case 5) is increased compared with the reference state, the FS response tends to separate appreciably from the SS of the structure. Moreover, looking at the case where C_M and C_R are converged at one single point but dislocated from the C_G of the structure, it can be seen that there is no appreciable difference between the FS response and the SS response under the direction of major component of seismic excitation. However, the response difference between the two edges is appreciably large under minor component of seismic excitation. In terms of response variation along the height of the structure for the same case of asymmetry, it can be seen that the influence is high at the top roof level compared with lower floors of the structure. This validates the fact stated earlier in regards to the global behavior of structures that top floors experience the highest influence of asymmetry when the planar eccentricities are uniformly distributed along the height of the structure. In terms of vertical mass eccentricities (case 7, case 8 and case 9), it can be seen that top floor's eccentricity has the maximum influence on the lower floor's response and lower floor eccentricity has the maximum influence on the top floor's response. When the eccentricities are on the intermediate floor, the influence is transmitted to the adjacent lower and top floors.

- In bi-eccentric TU-TS square structure with mass and stiffness eccentricities, top floor eccentricity has the highest influence on the maximum acceleration demand at the lower floor levels. Conversely, lower floor eccentricity has the highest influence on the top floor's acceleration demand.
- In bi-eccentric TU-TS square structure with mass and stiffness eccentricities, the top floor experiences the highest influence of asymmetry when the planar eccentricities are evenly distributed in vertical direction of the structure.
- Eccentricities at a floor tend to transmit their influence to the adjacent floors.

6.3.3 Acceleration response of mono-eccentric S-1 model

In the case of the mono-eccentric S-1 model, it can be seen in Figure 6.3 that the acceleration demands at the two edges have shown similar response trends under near-field and far-field seismic excitation. Unlike angular drifts, acceleration demands have been found to be higher at the FS compared with the SS of the structure for all cases of asymmetries. In this case, Y-direction is the direction of eccentricity. In particular, from case 4 and case 7, it can be seen that the top floor eccentricity has the highest influence on maximum acceleration demands at the first-floor level for both near-field and far-field seismic excitation. On the other hand, first-floor eccentricities (case 2 and case 9) have the highest influence on the top floor's acceleration demands under both near-field and far-field seismic excitation. Furthermore, it can be seen that this influence is equally sensitive at the FS and SS of the structure.

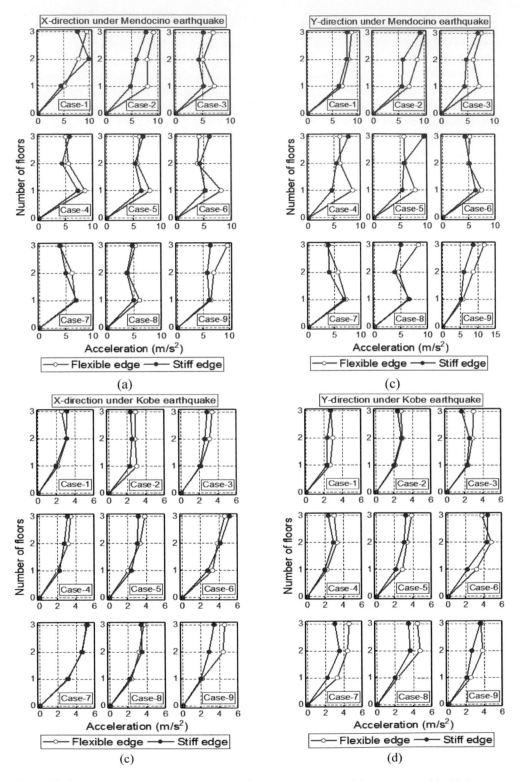

Figure 6.3 Enhanced acceleration response of mono-symmetric S-1 structure: (a) X-direction response under Mendocino earthquake, (b) Y-direction response under Mendocino earthquake, (c) X-direction response under Kobe earthquake and (d) Y-direction response under Kobe earthquake

Differentiating between the seismic demands of RI (case 1 and case 5) and IRI scenarios (all cases except case 1 and case 5), it can be seen that response transition from the FS towards the SS is widespread in the direction of eccentricity at second- and third-floor levels. This in turn gives rise to the fact that the top floor's response is more enhanced and the response transition from the FS towards the SS is highly evident in the direction of eccentricity if all the floors carry approximately equal amount of planar eccentricity along the height of the structure. Furthermore, it can be seen that increasing the intensity of eccentricity (case 5) compared with the reference state, no appreciable difference in the SS and FS response at first-floor level is observed. However, moving towards the upper floors, the seismic demands considerably change at the FS and SS along the direction of eccentricity. Looking at the case where C_M and C_R are converged at one single point but not at the C_G of the structure, it can be seen that the response transition between the FS and SS is highly evident in the direction of eccentricity. Conversely, the response transition is negligible in the orthogonal direction. It is interesting to note that seismic demands under such scenarios may likely decrease from the reference state of the irregular structure. This response reduction is higher at the SS compared with the response at the FS of the structure. In terms of response variation along the height of the structure for the same case of asymmetry, it can be seen that the influence is high at the top floor level compared with lower floors of the structure. This validates the fact stated earlier that the top floor experiences the highest influence of asymmetry when the planar eccentricities are uniformly distributed along the height of the structure. In terms of vertical mass eccentricities (case 7, case 8 and case 9), it can be seen that the top floor's eccentricity has the maximum influence on the lower floor's response and lower floor eccentricity has the maximum influence on the top floor's response. When the eccentricity is in the mid-floor level, it transmits its influence to the adjacent consecutive floors.

- In mono-eccentric TU-TS-square structures with mass and stiffness eccentricities, top floor eccentricity has the highest influence on the maximum acceleration demand at the lower floor levels. Conversely, lower floor eccentricity has the highest influence on the top floor's acceleration demand. However, this influence is dominant only in the direction of eccentricity.
- In mono-eccentric TU-TS square structures with mass and stiffness eccentricities, the top floor experiences the highest influence of asymmetry when the planar eccentricities are uniformly distributed along the height of the structure. This can be observed from appreciably variant seismic response at the FS and SS of the structure in such kind of scenarios.
- In mono-eccentric TU-TS structures, maximum influence on the seismic demands tends to occur in the direction of eccentricity. Seismic demands corresponding to both the FS and SS of the structure for all cases of eccentricities are appreciably different from each other in the direction of eccentricity. For the orthogonal direction, no appreciable difference in the SS and FS response is observed. This leads to the fact that in such kind of structures the torsional influence is dominant along the direction of eccentricity only.
- Regardless of the direction of eccentricity and type of seismic excitation, first-floor response is the least influenced by torsional vibrations and produces similar seismic demands at both the FS and SS of the structure.
- The top floor's response is the most influenced regardless of the location of eccentricity.
- The location of eccentricity plays a pivotal role in determining the behavior of the structure. Lower floor eccentricities tend to transmit its influence to the top floor. The middle

floor, on the other hand, transmits its influence to its consecutive adjacent lower and top floor altogether while the middle floor's response remains unaffected.

- Regardless of the direction of eccentricity, seismic demands may likely decrease at both the FS and SS of the structure under case 6 (when C_M and C_R are converged but not located at the C_G of the structure) compared with the RI state. Response reduction is higher at the SS of the structure compared with the FS response.
- In terms of vertical mass eccentricities, it can be seen that top floor's eccentricity has the maximum influence on the lower floor's response and lower floor eccentricity has the maximum influence on the top floor's response with negligible influence on the response transition from the FS to the SS or from the SS to the FS of the structure.

6.3.4 Acceleration response of S-2 model

In the case of S-2, it can be seen in Figure 6.4 that the acceleration demands at the two edges and for the selected excitations have considerably abnormal trends. Seismic demands are higher at the SS compared with the FS of the structure under Kobe earthquake. Conversely, the demands are relatively higher at the FS of the structure under Mendocino earthquake. This leads to the fact that the ground motion sensitive response may lead to abnormal response transition between the FS and SS of the structure. In particular, from case 4 and case 7, it can be seen that the response transition between the FS and SS is relatively negligible under both Mendocino and Kobe earthquakes for case 4 while the response transition is appreciably large for case 7. Moreover, it can be seen that both the edges have induced higher seismic demands at the first floor despite the fact that vertical mass eccentricity is located at the top floor of the structure. This leads to the conclusion that the top floor's vertical mass eccentricity transmits its influence to the adjacent lower floors with highest influence being on the first-floor level. On the other hand, first-floor eccentricities (case 2 and case 9) have the highest influence on the top floor's acceleration demands under Mendocino earthquake. However, an entirely opposite trend in the seismic demands can be observed under Kobe earthquake, where maximum response corresponds to the location of eccentricity and therefore, acceleration demands are higher at the first-floor level with a sudden reduction in the seismic demands at the adjacent floors. Differentiating between the seismic demands of case 1 and case 5, it can be seen that response transition from the FS towards the SS is widespread at the top floor level under Mendocino earthquake while the response transition is widespread at the first-floor level under Kobe earthquake. Similarly, seismic demands are high at the FS under Mendocino earthquake while the demands under Kobe earthquake are higher at the SS of the structure. From the response of case 1, it can be said that regardless of the seismic excitation, the torsional influence is only high at the top floor level while other floors remain unaffected. Looking at the case where C_M and C_R are converged at one single point but not at the C_G of the structure, it can be seen that the response transition between the FS and SS is highly evident under Kobe earthquake. The interesting fact here is to note that shifting both C_V and C_R from the C_G of the structure transfers its influence to the first-floor level as opposed to the case when both C_V and C_R are at the C_G of the structure and torsional influence is high only at the top floor level (case 1). In terms of vertical mass eccentricities (case 7, case 8 and case 9), it can be seen that under Kobe earthquake, the torsional influence at the first-floor level is significantly high when the vertical mass eccentricity is at the top floor level. Moreover, this torsional influence further aggravates as soon as the vertical mass eccentricity moves down the floor. While torsional influence in terms of response

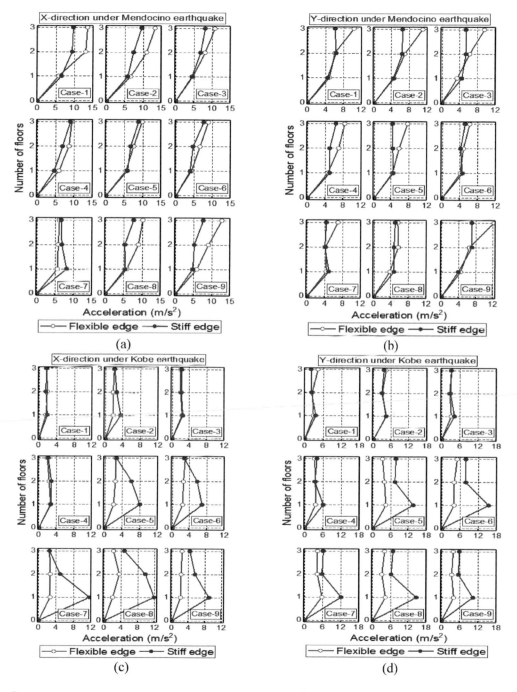

Figure 6.4 Enhanced acceleration response of S-2 structure: (a) X-direction response under Mendocino earthquake, (b) Y-direction response under Mendocino earthquake, (c) X-direction response under Kobe earthquake and (d) Y-direction response under Kobe earthquake

transition from the FS towards the SS is influenced to a great extent, it is surprising to note that in terms of absolute translational response, the FS of the structure remains unaffected at all floor levels. As opposed to the Kobe earthquake, seismic response under the Mendocino earthquake has demonstrated an entirely opposite trend where the location of eccentricity has influenced the seismic response. In a nutshell, higher influence at the top floor level under the top floor's vertical mass eccentricity is observed. Similarly, higher influence at the first-floor level under the first floor's vertical mass eccentricity is observed.

- In a bi-eccentric TU-TF square structure with mass, stiffness and strength eccentricities, the top floor's vertical mass eccentricity has the highest influence on the maximum acceleration demand at the adjacent lower floor levels with the highest influence being on the first-floor level.
- The response transition from the FS towards the SS or from the SS towards the FS becomes abnormal under seismic excitation sensitive to such kind of structures and leads to significantly higher seismic demands at the SS of the structure unlike the other seismic excitations where maximum acceleration demands occurred at the FS of the structure.
- Response augmentation at the SS under the vertical top floor mass eccentricity against Kobe earthquake is almost 3 times at the adjacent second floor and 6 times at the first-floor level in the orthogonal direction. Conversely, the response augmentation at the flexible edge is almost negligible. Surprisingly, the response at the top floor level reduces under both Kobe and Mendocino earthquakes in the orthogonal directions compared with its RI state.
- Maximum seismic demands are influenced by the location of eccentricity in such kind of structures under the excitation sensitive to the structure. Unlike other structures, the SS of the structure may likely produce higher seismic demands at the floor having eccentricities. Under this kind of scenario, the response augmentation is 4 times at the SS of the structure while the FS of the structure may remain unaffected.
- In the RI of the structure, regardless of the type of seismic excitation, the torsional influence is only high at the top floor level while other floors remain unaffected leading to the conclusion that in strength and stiffness eccentric structures, the highest torsional influence corresponds to the top floor level (De-La-Colina, 1999a, 1999b). This conclusion becomes invalid when C_M and C_S are converged at one single point but dislocated from the C_G of the structure. Under this condition, the first-floor response is greatly influenced while the top floor remains the least affected.
- In such kind of structures, higher torsional influence may occur at the first-floor level in terms of response transition from the FS towards the SS. In terms of absolute translational response, the FS remains entirely unaffected under all cases of vertical mass eccentricity. It should be noted that this particular observation is only associated to the seismic excitation sensitive to the mass, stiffness and strength eccentric structure.

6.3.5   Acceleration response of S-3 model

In the case of the S-3 model, it can be seen in Figure 6.5 that the acceleration demands at the two edges and for the selected excitations have approximately similar response trends. For all cases of irregularities, maximum response only occurs at the FS compared with the SS of the structure with few exceptions. For case 4 and case 7, it can be seen that the

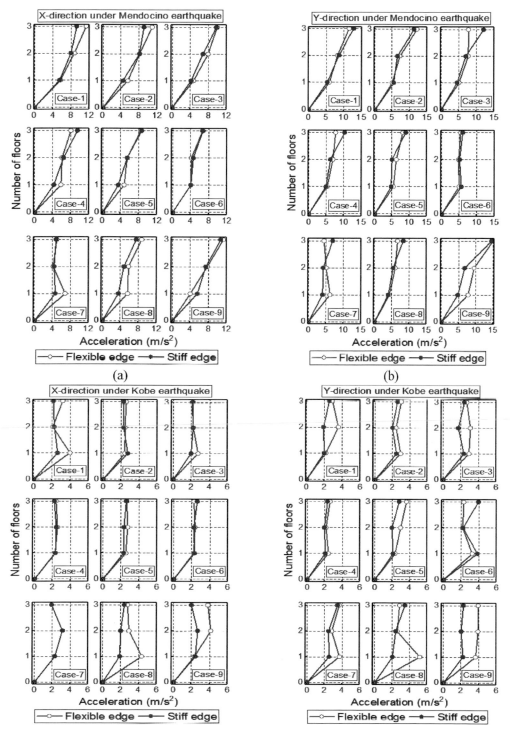

Figure 6.5 Enhanced acceleration response of S-3 structure: (a) X-direction response under Mendocino earthquake, (b) Y-direction response under Mendocino earthquake, (c) X-direction response under Kobe earthquake and (d) Y-direction response under Kobe earthquake

response transition between the FS and SS is relatively negligible under both Mendocino and Kobe earthquakes which demonstrates negligible torsional influence under such conditions. Moreover, it can be seen that both the edges have induced relatively higher seismic demands at the first floor despite the fact that planar and vertical mass eccentricity is located at the top floor of the structure in these cases. This leads to the conclusion that the top floor's planar and vertical mass eccentricity in such kind of structures transmit its influence to the adjacent lower floors with highest influence being on the first-floor level regardless of the type of seismic excitation. On the other hand, first-floor eccentricities (case 2 and case 9) have the highest influence on the top floor's acceleration demands under both Kobe and Mendocino earthquakes with considerable influence of torsional dominance in the seismic response in terms of response transition from the FS towards the SS of the structure. This response transition is appreciably large at the top floor level and considerably negligible at the first-floor level. Differentiating between the seismic demands of case 1 and case 5, it can be seen that response transition from the FS towards the SS is negligible at all floor levels with few exceptions for both Kobe and Mendocino earthquakes. For the case when C_M and C_R are converged at one single point but not located at the C_G of the structure, it can be seen that the influence of seismic response is generally higher at the first-floor level. As observed in the other structures, it can be seen that torsional influence is negligible in such kind of asymmetric case. In terms of vertical mass eccentricities (case 7, case 8 and case 9), higher torsional influence at the first-floor level can be observed when the vertical mass eccentricity is located in the lower floors. As the location of vertical mass eccentricity is shifted to the adjacent upper floors, the torsional influence seems to reduce. In terms of maximum response, it can be seen that eccentricity at the first floor induces higher response at the top floor level and influences the adjacent second floor's response as well. Similarly, the top floor's vertical mass eccentricity induces peak response at the first-floor level.

- In an S-3 structure with IRI state of irregularity, the top floor's vertical mass eccentricity has the highest influence on the maximum acceleration demand at the adjacent lower floor levels with the highest influence being on the first-floor level. Conversely, the lower floor vertical mass eccentricity has the highest influence on the top floor's response.
- Torsional response in terms of response transition from the FS towards the SS and from the SS towards the FS is relatively negligible in case of planar irregularities. However, this influence is considerably large when there are vertical mass eccentricities while the structure is originally in RI stiffness eccentric state.

6.3.6 Acceleration response of S-4 model

In the case of the S-4 model, it can be seen in Figure 6.6 that the acceleration demands at the two edges and for the selected excitations have approximately similar response trends. For all cases of irregularities, maximum response occurred only at the FS compared with the SS of the structure with few exceptions. For case 4 and case 7, it can be seen that the torsional influence in terms of response transition between the FS and SS is considerably high in the direction of major component of seismic excitation (X-direction for Mendocino earthquake and Y-direction for Kobe earthquake). Similarly, in terms of maximum response it can be seen that the seismic response is influenced at first-floor level under top floor's planar and vertical mass and stiffness eccentricities. Moreover, it can be seen that both the edges have

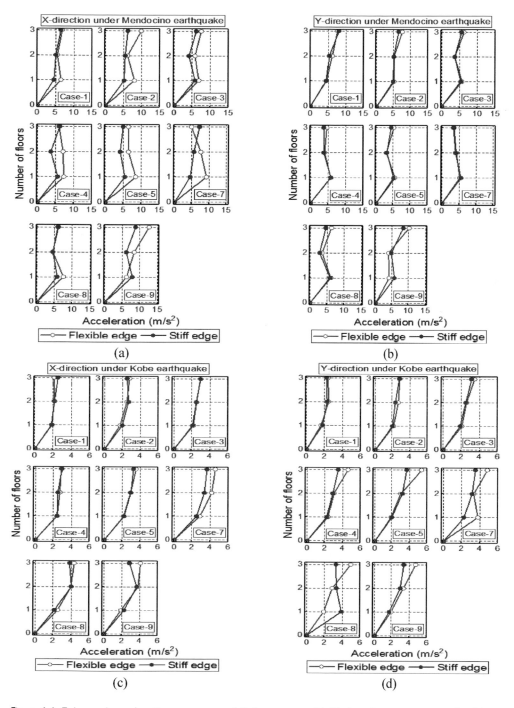

Figure 6.6 Enhanced acceleration response of S-4 structure: (a) X-direction response under Mendocino earthquake, (b) Y-direction response under Mendocino earthquake, (c) X-direction response under Kobe earthquake and (d) Y-direction response under Kobe earthquake

induced relatively higher seismic demands at the first floor despite the fact that planar and vertical mass eccentricity is located at the top floor of the structure in these cases. This leads to the conclusion that the top floor's planar and vertical mass eccentricity in these kinds of structures transmit its influence to the adjacent lower floors with highest influence being on the first-floor level regardless of the type of seismic excitation. On the other hand, first-floor eccentricities (case 2 and case 9) have the highest influence on the top floor's acceleration demands with considerable influence of torsional dominance in the seismic response in terms of response transition from the FS towards the SS of the structure in the direction of major component of seismic excitation. This response transition is appreciably large at the top and first-floor level and considerably negligible at the second-floor level. Moreover, it can be seen that first floor planar and vertical eccentricity has caused response reduction at the second-floor level in the orthogonal direction. Differentiating between the seismic demands of case 1 and case 5, it can be seen that the influence on the response is higher at the first-floor level for both Kobe and Mendocino earthquakes. For this model, case 6 has not been presented due to some disruptions in the obtained data. In terms of vertical mass eccentricities (case v7, case 8 and case 9), higher torsional influence at the first and top floor levels can be observed. These cases lead to the fact that lower floor vertical mass eccentricity has the highest influence on the top floor while the top floor has the highest influence on the first-floor level. When the vertical mass eccentricity is at the middle floor, its corresponding influence is transmitted to the adjacent lower and upper floors.

• In an S-4 structure with an IRI state of irregularity in its reference state, the top floor's vertical mass eccentricity has the highest influence on the maximum acceleration demand at the adjacent lower floor levels with highest influence being on the first-floor level. Conversely, the lower floor's planar and vertical mass eccentricity has the highest influence on the top floor's response.
• Eccentricity at a floor may likely cause response reduction at the adjacent floor level.
• Influence of torsional vibration is more dominant in the direction of major component of seismic excitations.

6.4   Deformation response

6.4.1   Deformation response of RC model

The acceleration signals at each point of the model measured by each ground motion were subjected to baseline correction and filtering and then the two points were obtained from the displacement time curve. In Figure 6.7a, the maximum displacement at the centroid has been presented. Since the stiffness edge is limited to the region near the structural shear wall, the rest of the part of the structure is termed as FS as the C_R in this case is very close to the shear wall. The global displacement responses obtained at C_M correspond to the FS and represent the behavior of structure at the FS. It can be seen that the displacement responses were gradually increasing from bottom to top under progressive seismic excitations. It can also be observed that the seismic response of the experimental structure in different experimental stages was mainly based on the basic modes with some contribution of variation in structural dynamic properties under progressive increase in the input ground acceleration. The overall lateral displacement characteristics are obvious in Figure 6.7a and the results correspond to the dynamic response curves and the basic modes shear characteristics of the structure.

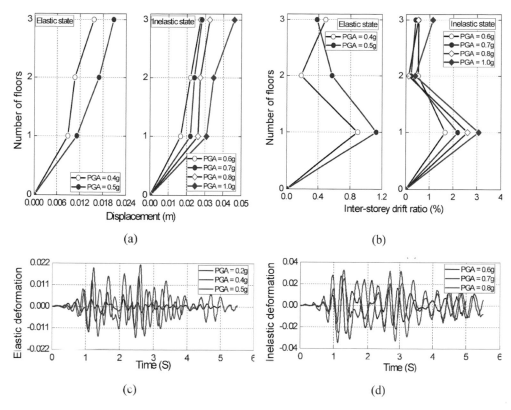

Figure 6.7 Elastic and inelastic deformation response of RC model: (a) floor displacements, (b) inter-story drift ratios, (c) top-roof displacement history response in the elastic state and (d) top-roof displacement history response in the inelastic state

In Figure 6.7a, it can be seen that for seismic loads with PGAs from 0.1g to 0.3g, the structural response was less sensitive to damage due to structure being intact in the elastic range. Thereafter, the damage developed progressively after the PGA of input ground motion was increased from 0.3g to 0.5g. Due to the damage induced in the structure, the structure's dynamic properties were greatly compromised and varying structural dynamic properties caused change in the response of the structure as well. For PGAs ranging from 0.4g to 0.6g, the displacement response pattern appears to be the same. However, the progressive increase in the input ground acceleration from 0.7g to 1g caused variation in the displacement response pattern of the structure which is again due to the non-uniform yielding and thus cracking, which ultimately has influenced the dynamic properties of the structure (Fig. 6.7a, c and d).

To explore the contribution of inelastic deformation further along the height of the structure, Figure 6.7b demonstrates the response of the inter-story drift ratios (IDR) measured under progressive seismic excitations. The dominance of the first mode can easily be observed. Nevertheless, the structure showed pretty much observable crimp in the deformed shape at the first story, demonstrating an obvious higher-mode involvement. This observable crimp at the first story for PGA of 0.6g–1.0g demonstrated the abnormal influences which occurred due to change in the dimensions of the structural components under progressively increased loading. This eventually influenced story stiffness due to the formation of damage and variation in size of structural components at different locations in the structure. The structure deteriorated rapidly under the influence of asymmetry for higher PGAs and reached the advanced inelastic response with an illustration of more critical drift response at the first story approaching 1.5% under PGA = 0.6g. Since in Figure 6.7b the sudden change in the displacement demand after the first story indicates the influence of higher mode effects and dynamic properties of the structure which kept on changing due to the formation of cracks in the structure, it can also be said that at lower PGAs (0.4g–0.5g), the sudden change in the IDR demand was solely influenced by the higher mode effects, while at higher PGAs (0.6g–1.0g) this change was mostly caused by structural dynamic properties variation as the story stiffness during this stage was greatly compromised. Nevertheless, acceleration responses in contrast to the displacement responses appear to have less sensitivity towards local oscillations. This happened due to the fact that low-frequency global vibrations, caused by the input excitation, dominated the main wave-form and caused the local oscillations to produce additional acceleration of high frequency with local effects. It can be said that in damage identification studies, suddenly varying patterns of acceleration responses can be adopted as one of the indexes for identification purposes.

6.4.2 Angular drift response of steel models

Figures 6.8–6.9 illustrate the envelope of the angular drift response at the FS and SS of the structure in the orthogonal directions under bi-directional seismic excitations. The responses have been normalized with the angular drift response at the C_G of the structure. Based on the detailed results in Figures 6.8–6.9, the estimation of torsional effects can be seen in these envelopes at both the FS and SS. It can be seen that in TF structures, the angular drift at the SS has increased, compared to those in the corresponding FS of the TF structures. In general, typical quantitative behavior of TS and FS is presented in these figures which show the variations of lateral angular drift demands at both the FS and SS, with respect to their counter TB structure. Figures 6.8–6.9 illustrate the torsional behavior of all structural models with varying eccentricities, but with different torsional stiffness and strengths. Presented is the peak angular drift response at the top of the buildings during bi-directional seismic excitation of the TU structures. These figures are helpful in the understanding of the seismic response magnification at the SS and FS in both TS and TF structures. Note, however, that in these figures, the angular drift response is a straight horizontal line for TU structures. Therefore, the response at both the FS and SS in all the TU structures is representative of response magnification compared to that of TB structures.

In case of bi-eccentric S-1 model, it can be seen that overall angular drift response is greater at the FS compared with the SS response for all types of seismic excitations and for all cases of eccentricities except Y-direction under far-field Kobe earthquake where response under case 6 is greater at SS compared with the FS response. The same case of eccentricity in the Y-direction under near-field Mendocino earthquake has demonstrated entirely different results compared with far-field seismic excitation. This is due to the fact that near-field seismic excitation has the dominant seismic component along the X-direction whereas in the far-field seismic excitation, the dominant seismic component is along the Y-direction. It is interesting to discuss the scenario of asymmetry where C_M and C_S are converged at one single point but dislocated from the G_C; it can be seen that angular drift response at the SS under this scenario is greater than the response at the FS under near-field seismic excitation. Conversely, an entirely different response is observed for the two edges under far-field seismic excitation where the FS demonstrated greater response compared with the SS. In terms of vertical mass irregularity with an IRI state, regardless of the trend of maximum response at the FS and SS, it can be seen that under both Mendocino and Kobe earthquakes, top floor vertical mass eccentricity has caused maximum influence on the top floor angular drift response. In the similar fashion, second-floor vertical eccentricity has caused maximum response top floor angular drift response under the Mendocino earthquake as opposed to the Kobe earthquake where the maximum top floor angular drift response is observed under vertical mass irregularity at the top roof level. The overall observations can be summarized as follows:

- In terms of angular drift of S-1 structure, lower floor vertical mass irregularity has the highest influence at the top floor under near-field seismic excitation whereas higher floor vertical mass irregularity has the highest influence at the top floor under the far-field seismic excitation.
- Angular drift response is entirely abnormal when the C_M and C_S are converged but dislocated from the C_G of the structure.
- Top floor mass and stiffness eccentricities in the reference state have shown the highest influence on the angular drift response among all types of investigated asymmetries.

In the case of the mono-symmetric S-1 model, it can be seen that the rotational drift response at the two edges has not shown similar response pattern for both the Mendocino and Kobe earthquakes. This endorses the fact (which is presented in the next chapter) that seismic response is influenced by the characteristics of seismic excitation and the excited mode of the structure. Moreover, it can be seen that angular drift response at the SS of the structure is more sensitive to the Mendocino earthquake in the Y-direction compared with the Kobe earthquake. Comparing the peak response under near-field seismic excitation at the FS and SS, it can be seen that the majority of the asymmetric cases at the FS have demonstrated lower demands at the SS in the direction of eccentricity. Similarly, comparing the peak response under Mendocino earthquake at the FS and SS, it can be seen that the majority of the asymmetric cases at the FS have demonstrated simultaneously greater demands compared with the SS demands. It is also interesting to note that in case 8, peak angular drift response in the orthogonal directions under the Mendocino earthquake has demonstrated maximum demand among all cases of eccentricities. However, for the similar case of eccentricity under the Kobe earthquake, this observation was recorded only in the direction of eccentricity. Peak response in the corresponding orthogonal direction is governed by case 6. For synthesis of results, the overall observations can be summarized as follows:

- Seismic response is influenced by the characteristics of seismic excitation and the excited mode of the structure.
- Comparing the response of the two edges, it can be concluded that under near-fault seismic excitation (Mendocino earthquake), response is maximum at the SS in the direction of eccentricity whereas the response is maximum for most of the asymmetric cases at the FS in the orthogonal direction. Conversely, under far-field seismic excitation (Kobe

earthquake), the response is greater at the FS simultaneously in the orthogonal directions for the majority of the asymmetric cases.
- Peak response is sensitive to near-field seismic excitation simultaneously in the orthogonal direction even in the mono-eccentric S-1 model. On the other hand, under far-field seismic excitation, peak response is sensitive to far-field seismic excitation only in the direction of eccentricity.

In case of the S-2 structure, it can be seen that the rotational drift response at the two edges has shown somewhat similar trends for both near-field and far-field seismic excitations. It should be noted that this structure is an IRI structure even in its reference state with bi-eccentric mass, stiffness and strength eccentricities. It is interesting to note here that in such kind of TU structures, vertical mass irregularity at the first-floor level tends to influence the maximum response at the third-floor level. Case 9, where the mass of the first floor is 3 times the mass of the consecutive upper floors, has the highest influence among all kinds of asymmetric scenarios for this kind of structure. In general, angular drift response was found to be higher at the SS compared with the FS of the structure. The overall observations can be summarized as follows:

- In an S-2 structure having IRI state of irregularity with mass, stiffness and strength eccentricities, lower floor asymmetry tends to influence the maximum top floor angular drift the most compared with all other cases of asymmetries.

In an S-3 structure, it can be seen that the rotational drift response at the two edges has demonstrated an entirely different trend for both near-field and far-field seismic excitations. Focusing specifically on the scenario where C_M and C_S are converged but dislocated from the G_C of the structure, near-field seismic excitation and far-field seismic excitations have presented entirely different trends. Under near-field seismic excitation, the SS tends to produce greater response compared with the FS whereas in the case of far-field seismic excitation, the FS has produced greater angular drift demands compared with the SS of the structure. Specifically looking at the IRI state of the structure in terms of vertical mass irregularities, it can be seen that the top floor eccentricity has the maximum influence on the top floor's response. The overall observations can be summarized as follows:

- In an S-3 structure, IRI state of irregularity with mass and stiffness eccentricities at the top roof level, the angular drift response has the maximum influence on the rotational response.
- Lower floor eccentricity has the maximum influence on the maximum angular drift response of top story level along the direction of minor seismic component.

In an S-4 structure, it can be seen that the rotational drift response at the two edges has shown an entirely different trend under near-field and far-field seismic excitation as has been discussed regarding S-3 structures as well. This in turn leads to the conclusion that angular drift response is more complex in pattern for TU-TF structures as opposed to the case of TU-TS structures where angular drift response has demonstrated a fairly predictable trend at the FS and SS. Looking at the angular drift response for the scenario where C_M and C_S are converged at one single point but dislocated from the C_G of the structure, it can be

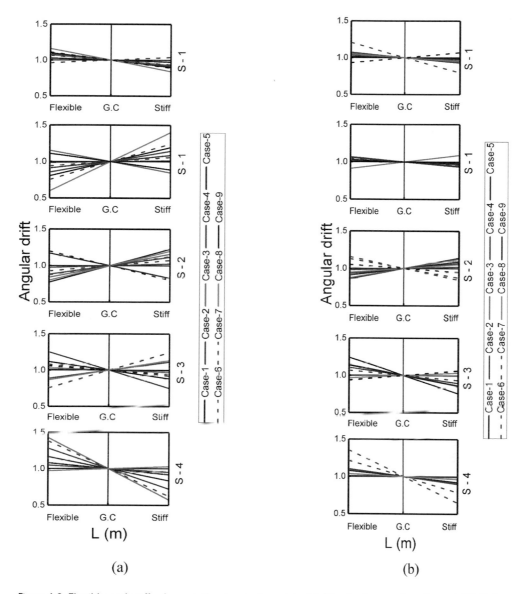

Figure 6.8 Flexible and stiff edge rotation in steel models: (a) Mendocino earthquake and (b) Kobe earthquake

seen that this scenario has the highest influence on the angular drift response under both near-field and far-field seismic excitation with a fairly predictable pattern of the response at the FS and SS. Furthermore, it can be seen that the highest angular drift response was recorded at the FS compared with the SS of the structure. This validates the conclusion for the TU-TF structure that the highest angular drift response occurs at the FS of the structure in terms of case 6. Vertical mass irregularity at the first floor has demonstrated highest influence not only at the top floor's angular drift response but has proved to be the worst scenario among all eccentricity cases. This fact is validated in all IRI state of irregularities where it can be seen that the asymmetric influence has produced the highest top floor's angular drift response with few exceptions. Considering the response transition from the FS towards the SS for the same IRI cases under near-field and far-field seismic excitations, it can be seen that angular drift response is greater at the FS of the structure compared with the SS of the structure.

• In an S-4 structure having an IRI state with mass, stiffness and strength eccentricities, all IRI states tend to cause maximum influence at the top floor level in terms of angular drift response.

6.5   Discussion on the global damage behavior

6.5.1   Global behavior of RC model

Figure 6.9 illustrates the system-level twist impacts on the behavior of the asymmetric shear wall structure, where the structure experiences significantly different twist demands throughout the structure's seismic input history. Interaction between the induced displacement demands imposed on the structure and the structural twist demands is responsible for different levels of twist demands. Frame-shear wall asymmetrical structure had an induced eccentricity of 0.45 times the longitudinal direction, eventually causing the twisting effect under translation ground motion. The structure's center of resistance is much closer to the SS; therefore, rotational effect at the SS is approximately negligible (Fig. 6.9a). Therefore, maximum torsional responses do not correspond to the SS; neither do the simultaneous maximum displacements. This gives rise to the need to evaluate the locations of center of rigidity and building-level twists throughout the loading history. Utilizing the floor slab's displacement responses, the drift angles (Fig. 6.9b and c) were investigated throughout the story height. The floor drift angles and their corresponding inter-story drift angles were investigated. Observing the inter-story drift angles in combination with Figure 6.7b, a gradual increase in the inter-story drift angle with the increasing level of input seismic excitations up to the peak capacity of the structure (corresponding to IDR of approximately 1.5%) can be observed. At that PGA level of 0.6g, which was accompanied by mild damage in the structure, significant rise in the floor twist occurred and almost caused an increase of 50% increase in the inter-story drift angles when strength degradation started during an input excitation of 0.7g (Fig. 6.9c). One interesting fact can also be observed that the increase in the IDRs at the first floor has caused a decrease in the inter-story drift angles at the first floor. Looking for the same fact at the second floor, it can be seen that the decrease in the IDRs has caused an increase in the inter-story drift angle at the second floor. The IDRs again started to increase with a kink followed by a decrease in the inter-story drift angles. This behavior remained equally constant both in the elastic and inelastic state. Furthermore, it can be observed that sudden change in the demand occurred both at the first and second floor in case of IDRs and inter-story drift angles. In order to determine a trend between the IDRs and IDR-angles, it can be said that IDRs are apparently inversely related to the inter-story drift angles. This trend was almost identical both in the elastic and inelastic state. Looking at the height-wise profile of IDR-angles, it can be seen that the IDR-angles increased until the second floor, with a sudden change in the demand after second floor.

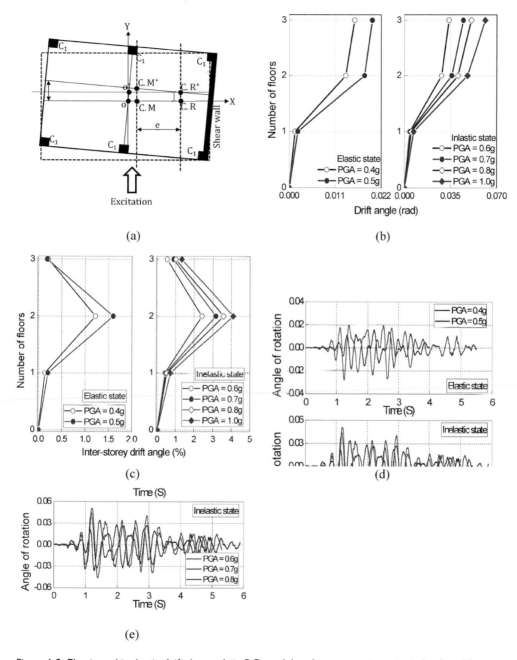

Figure 6.9 Elastic and inelastic drift demands in RC model under progressive seismic loading: (a) structural floor rotation through center of resistance, (b) drift angles, (c) inter-story drift angles, (d) angle of rotation in shear wall in the elastic state and (e) angle of rotation in shear wall in the inelastic state

This sudden change in the demand was contributed by the influence of torsional mode. It can be observed from Figure 6.9b and c that the structure's rotational response was asymmetrical along the height of the structure, indicating obvious asymmetric twist characteristics in loading direction.

When elastic torsional response of a structure is under consideration, only the center of rotation/rigidity (C_R) and center of mass (C_M) locations are of interest but when a structure responds within its inelastic state, which is expected under progressive seismic loads, the center of strength (C_V) is as important as the other locations in causing the twist to the structure. Therefore, it can be said that three important locations played a vital role in bringing the twist response of the structure, namely C_M, C_R and C_V. In the inelastic range, shear wall stiffness is likely to change due to transformation from an elastic to a highly inelastic state. The subsequent influence of this rotation would be on the location of C_V and C_R, and the corresponding redistribution of the load followed by the damage in the stiff elements (Heerema *et al.*, 2014). Since the contribution of the shear wall towards lateral resistance is very high in the considered experimental structure, it can be seen that stiffness reduction is approximately negligible both in the elastic and inelastic state (Fig. 6.9d and e). Therefore, it can be said that in this particular case, the change in the locations of C_V and C_R is approximately negligible during the transformation of the structure form elastic to inelastic state.

Since the shear wall is aligned along the direction of input excitation, the contribution of the shear wall towards resistance was about 96% throughout the experiment. Since the shear wall of the structure did not form any damage along with negligible twist at the SS, resulting damage and twist eventually occurred at the FS (Fig. 6.9a). Once the significant damage developed, that is when the structure entered into the inelastic state, and higher twist angles developed at the FS. The reason behind the relatively higher twist response at the FS is the torsional stiffness degradation. Figure 6.9d and e provide an insight into the contribution of shear wall towards resistance along the direction of seismic excitation. The figures demonstrate that the role of shear wall towards the structure's resistance was essentially constant throughout the loading history for both elastic and in the inelastic states. This attributes to approximately negligible degradation in the strength and higher displacement capacity at the SS of the structure.

The structure's ductility demands were observed in terms of ductility ratio μ and lateral stiffness ratio η_Δ and were plotted against normalized weight of the structure α as shown in Figure 6.10. The ductility ratio μ corresponds to the ratio of inelastic displacement to the elastic displacement in the perfectly elastic state while the ductility ratio μ_m corresponds to the ratio of inelastic displacement to the elastic displacement in the micro-cracking state. The lateral stiffness ratios were computed as the ratio of story drifts to the mean story drifts. The normalized weight of the structure for an ith story is considered as the ratio of weight above the ith floor to total weight of the structure. The ductility demands were then plotted against the normalized weight to observe the influence of progressive seismic loading on structural ductility demands. From Figure 6.10a, it can be seen that ductility ratios under progressive loading remained considerably uniform with a uniform increase in the demand under the increase of input seismic excitation, when compared both with the completely elastic state and micro-cracking state. On the contrary, it can be seen in Figure 6.10b that increase in the PGA didn't cause significant stiffness degradation except in the first floor where the stiffness reduction is quite obvious.

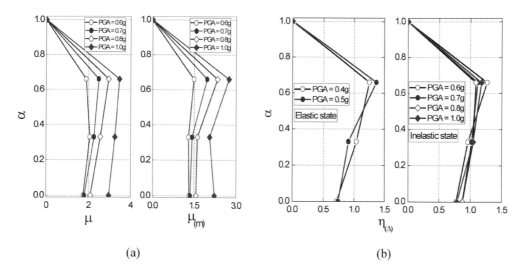

Figure 6.10 Ductility demands in RC model: (a) ductility ratio and (b) lateral stiffness ratio

6.5.2 Correlating dynamic properties of RC model with global response

Figures 6.11 and 6.12 represent the dynamic properties of the structure under progressive loading. During the elastic phase, the first order mode shape of the test structure was between the bending deformation and shear type deformation. At this point the overall structure and the inter-story stiffness were large and hence the lateral shift was closer. From Figure 6.11a and b, it can be seen that the first and second mode damping and the frequency changes were not obvious in the elastic state. However, once the model was cracked, the first and second order frequencies decreased gradually and the corresponding damping ratio increased simultaneously. The floor stiffness decreased gradually which eventually caused an increase in the inter-story deformation. The augmentation of the damage in the structure led to the increase in the width of beams and columns; therefore, the increase in the length of the structural members caused significant and rapid reduction in the frequency. However, this phenomenon consequently caused significant and rapid rise in the damping ratio of the structure.

From the second-order frequency reduction point of view, the second-order frequency reduction is comparatively the same as first-order frequency reduction under progressive increase in the ground motion excitation. Similar observations have been recorded for the damping ratio. Frequency degradation corresponds to the cracking in the structure beforehand. The structural internal cracking was first analyzed using the dynamic properties of the structure. From Figure 6.11, it can be seen that the last degradation stage corresponds to significant damage in the structure, that is, when plastic hinges started to form in the structure. But it does not mean that the structure collapsed; however, it definitely means that the structure has reached the inelastic state. Indeed, this cracking is responsible for the frequency shift and change in the dynamic properties of the structure.

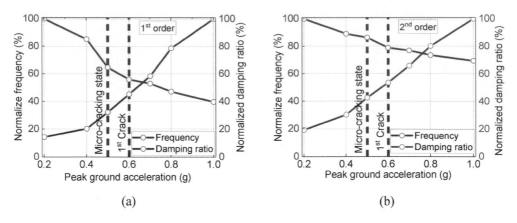

Figure 6.11 Dynamic properties of the structure: (a) frequency development and (b) damping ratio development

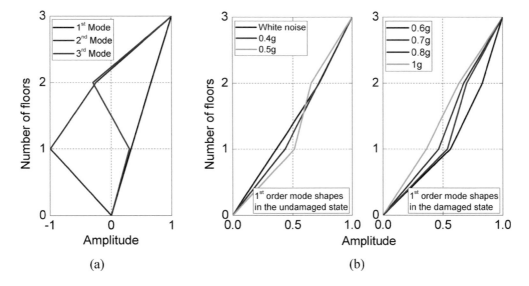

Figure 6.12 Structural dynamic modes under progressive loading: (a) undeformed structure and (b) deformed structure

6.5.3   *Discussion on the global response of steel models*

In the case of TS structures, the rotations are moderate at the SS of the structure which also validates the findings of Fajfar *et al.* (2005). In general, for the case of synthesis, the difference in the translational response at opposite edges is relatively smaller under all seismic excitations compared with rotational demands. For rotational demands corresponding to the FS, the demands are not as predictable as in the case of the SS in both TS and TF structures. In general, when mass, stiffness and strength eccentricities are present, regardless of the characteristics of seismic excitations, larger rotational demands are observed. Rotations in the X-direction are somewhat stronger compared to Y-direction. However, the seismic response transition from TS to TF structure illustrates a complicated response transition especially for the FS of the structure. This is entirely opposite to the simulated local damage response presented in the previous chapters where damage response was more predictable for both the SS and FS. The complicated response transition in the global perspective has demonstrated several facts which for the general purpose of synthesis of results are mentioned as follows:

- Higher is the uncoupled torsional frequency ratio, lower is the angular drift response in all cases of irregularities and in all asymmetric cases of the presented models in this research corresponding to the mentioned criteria. Similarly, lower is the uncoupled torsional frequency ratio, higher is the angular drift response in all cases of irregularities and in all asymmetric cases of the presented models in this research corresponding to the mentioned criteria.
- Compared with the FS, case to case angular drift response variation is high at the SS of the structure in TS structures.
- For IRI state of irregularity, relatively equal case to case angular drift response variation was observed in TS systems. However, this trend cannot be observed in the case of Kobe earthquake where the SS response was lower than the FS response in all cases of asymmetries.
- Regardless of the fact that whether a structure is TS or TF, case 6 (when C_M and C_S are converged at one point but dislocated from C_G of the structure) has uniform influence on angular drift response corresponding to the FS and SS in all asymmetric structures. Moreover, these cases have demonstrated highest influence on the rotational demands compared with the translational demands.
- Regardless of whether a structure is TS or TF, case 7 (when the top floor mass is 3 times the mass of the consecutive lower floors with relatively equal static eccentricities at the intermediate and lower floors) has maximum influence on angular drift response corresponding to the FS and SS in all asymmetric structures.
- For IRI state of irregularity in terms of variation in the floor mass, maximum influence at the SS and FS occurs in all TS structures. Compared to the TS structures, this influence is relatively lower in TF structures at both the SS and FS.
- In TF structures, lower floor static eccentricities have the maximum influence on the angular drift response at the top floor level. Compared with TF structures, this influence is relatively low in TS structures.
- Angular drift response is equally high at the FS and SS of the structure in both TS and TF structures when the state of the structure is IRI compared with RI state of irregularity.
- Seismic response is strongly dependent upon the characteristics of seismic excitations and the excited mode of the structure.

6.6 Summary

The global behavior of various simple and complicated asymmetric systems, chosen to ensure wide applicability of results, is presented in this chapter with the following objectives:

- Investigating the effects of plan-wise distribution of stiffness, strength and mass on the seismic response.

- Evaluating the influence of vertical irregularities on a structure's global response.
- Investigating how the seismic response is influenced by the system parameters.
- Presenting how structural response is affected by horizontal and vertical asymmetry and how these effects differ between linear (elastic response for RC model and Kobe earthquake response for steel models) and nonlinear (inelastic response of RC model and El Centro, Mendocino and Northridge earthquake response for steel models) responses.
- Investigating the effects of the structural characteristics and the influence of the unidirectional (RC model) and bi-directional (steel models) seismic excitations on the global behavior of various asymmetric systems. Subsequently, the influence of asymmetry on system response is investigated by comparing the dynamic responses of asymmetric and symmetric systems.
- Investigating the characteristics of seismic excitation and the excited mode on the structural response of asymmetric structures.

For steel models, within the post-processing of the experimental results for each bi-directional seismic excitation, the seismic responses were normalized to those of the corresponding TB systems and their responses are presented and evaluated. Therefore, for the purpose of evaluation, the global responses have been demonstrated in terms of ratio of asymmetric response to symmetric response. This has been done to develop an understanding of response magnification in asymmetric structures. Presented first are the linear and nonlinear responses of asymmetric systems with varying distribution of stiffness/strength and mass. Based on these results, vertical mass eccentricity along with plan-wise distribution of stiffness and strength is shown to significantly affect the translational demands of asymmetric structures. It is demonstrated that nonlinear response, in particular strength-symmetric TS systems, is affected less by plan-asymmetry or system parameters compared to linear response. In general, for the sake of synthesis, it can be said that the linear response is less rotational than the nonlinear ones for TS structures. In case of TF structures, linear responses seem more rotational compared to the nonlinear ones.

6.6.1   Global behavior of C-1 model under torsional vibrations

In order to identify the ultimate state of collapse of a structure and structural performance, the horizontal displacement is primarily used as a threshold. Typical drifts estimated through various design-codes mostly come in the range of 1.5%–2.5% (EC-8, 2005; UBC, 1997). Previous research findings have even proposed a higher drift limit of up to 6% for actual failure of the structure based on simple structural member experiments (Roufaiel and Meyer, 1983). One other study (Lu, 2002) proposed inter-story drift limit of 3% and a drift limit of 2.5% at top story by performing experiments on different kind of structures and conducting a comparative study. The current experimental results for asymmetric frame-shear wall structure suggest that it is reasonable to expect a stable frame behavior if the following two conditions are met: (1) top-drift limit below 2.7%, and (2) inter-story drift limit below 1.5%. Hence, following are the observations that have been found in this research:

- The frame-shear wall structure maintained a stable behavior at a drift slightly lower than the proposed limits, as observed from the dynamic properties of the structure, i.e. PGA = 0.5g. However, the structure started inducing damage with further increase in the input ground motion. It can be said that the structure was transforming from a stable behavior to highly unstable behavior.

- The unstable behavior of the structure was first observed at PGA of 0.6g when the first-floor inter-story drift response was found higher and close to the proposed limit. The response was recorded in the form of a sharp increase of top displacement as well as the inter-story drifts. For such kind of irregular structures with local damage, it is understandable to lower the 2% top-drift limit. However, from a practical point of view, it doesn't seem to be important, as an inter-story drift of 1.5% is likely to govern in such kind of structures.

- The inter-story drift ratios and inter-story twist are inversely related to each other and provide a good measure of presence of asymmetry of the structure.

The proposed drift limits are limited to the structure which falls under the scope of similar asymmetric structures. The three-story asymmetric shear wall structure was exposed to progressively increasing ground motions. The following experimental observations have been made:

- Displacement responses are found to be less sensitive to local oscillations behaviors as compared to acceleration responses. This happened due to the fact that low-frequency global vibrations induced through the input excitation dominated the main wave-form and caused the local oscillations to produce additional acceleration of high frequency with local effects.

- Increase in the inter-story drift ratios at first floor has caused decrease in the inter-story drift angles at first floor. The similar inverse behavior was observed for other floors with identical trends both in the elastic and inelastic states. It is hypothesized that inter-story drift ratios are apparently inversely related to the inter-story twist.

Based on the structural damage response, top-story displacement and inter-story drift response, the top-drift has been proposed to be limited to 2% and the inter-story drift has been proposed to be limited to 1.5% as a general measure of stability for multi-story asymmetric frame-shear wall structures.

6.6.2   Global behavior of S-1 model under torsional vibrations

For S-1 structures, it can be said that regardless of the direction of eccentricity and primary component of seismic excitation, translational motion prevails. With reference to bi-eccentric S-1 structure, the maximum nonlinear acceleration responses are more uniform compared with the linear ones for all asymmetric cases, unlike mono-symmetric S-1 structures where linear responses are more uniform compared with the nonlinear ones for all asymmetric cases. However, it should be noted that with respect to mono-eccentric S-1 structures, the schemes with uni-directional eccentricity present greater linear and nonlinear seismic response variation and non-uniformity along the height of the structure only in the direction of eccentricity and under major component of seismic excitation. There is little to no variation in the corresponding perpendicular direction with no apparent eccentricity. Moreover, this is particularly evident on the FS of the structure. Similarly, IRI state of the S-1 structures has affected the bi-eccentric system the most compared with the mono-eccentric system. However, RI states of both mono-eccentric and bi-eccentric systems have shown similar trends.

6.6.3   Global behavior of S-2 model under torsional vibrations

In the case of S-2 structures, some more torsion can be seen. This is due to the fact that this structure is an IRI structure with strength, stiffness and mass eccentricity even in its reference state. However, stronger translational motion in both the X- and Y-direction is still evident. Why are global rotational demands less dominant than the translational demands? The simple answer to this question is because the system is a TS system. For the similar IRI case of S-4 structures, it was observed entirely the opposite as rotational demands were more evident in that system compared with the S-1 structure mainly because the S-4 structure is a TF structure.

6.6.4   Global behavior of S-3 model under torsional vibrations

For S-3 structures, the RI state of irregularity has the highest influence compared with the IRI state of irregularity. Moreover, in the IRI state of irregularity, the influence of torsional vibration is highest at the top floor level when the eccentricities are actually present in the lowest floor level. Similarly, lower floors are affected the most under torsional vibrations when the eccentricities are actually present at the top floor level.

In general, for the purpose of synthesis, it can be said that for S-3 structures, the influence of planar irregularity is negligible on the seismic response compared with the vertical irregularities.

6.6.5   Global behavior of S-4 model under torsional vibrations

The response of S-4 structures is predominantly torsional in the orthogonal directions. Moreover, the direction of the larger accelerations clearly corresponds to the direction of the stronger component of the ground motion. However, there is no clear correspondence of trend between translational response and the sense of rotation despite the fact that rotational response is more evident compared with the translation response. Since this structure is an IRI structure originally from its reference state, influence of torsional vibration is more dominant in the direction of major component of seismic excitations. Moreover, the second mode dominance is clearly visible in the seismic demands, and therefore, the seismic response is likely to get reduced at the intermediate floor level for all asymmetric cases.

Chapter 7

Influence of design parameters on the statistical distribution of structural response

7.1 Introduction

From the detailed investigation of the previous literature and analysis of the results in the previous chapters, it has become evident that seismic response is highly uncertain in nature because of its strong dependence on the following parameters:

- Excited mode of the structure.
- Type and location of inherent eccentricity of the structure.
- Ground motion characteristics.

The uncertain nature of the seismic response has been explained in detail in the previous chapter in the context of inherent eccentricities. The main focus of this chapter is the influence of ground motion characteristics on the seismic response of asymmetric structures. This chapter also brings attention towards the uncertainty of the observed damage study and the global behavior of the asymmetric structures in the previous chapters. In the context of varying orientations of seismic excitations used in this chapter, seismic demand significantly varies and thus indicates the strong uncertainty of the seismic response. Keeping in view the earlier described third parameter regarding the characteristics of ground motion, this chapter aims to provide a detailed insight into the seismic response uncertainty. For this purpose, a validated FE model in SAP2000 was developed and the global demands obtained from the C-1 model were validated with the FE model results.

7.2 Contribution of this chapter to knowledge

This chapter presents the influence of varying orientations of ground motions on C-1 model using a validated numerical model. For this purpose, experimental results were utilized for the assessment of theoretical and numerical results. The experimental observations were then compared with theoretical and numerical illustrations for each stage of seismic excitations from elastic to inelastic state of the structure for the validation of the theoretical and numerical response. Seismic responses at the FS and SS of the structure were compared to present the influence of asymmetry. Finally, the results were investigated from a statistical point of view to evaluate the response variability with varying orientations of seismic input. It has been found that:

- The SS of the structure is more sensitive to the rotational demands compared with the translational demands. This also validates the observations found in the previous chapter for rotational response of TS systems.
- The relative rotational response at the SS of a structure proved to have more response variability than at the FS of a structure under varying orientations of seismic excitation.

7.3   Varying orientations of seismic excitations

Most of the previous studies revolve around the estimation of maximum response through critical angle or through combination rules. However, to what extent the earthquake directionality can influence the performance of the structure is still a question. Also, the uncertainty of the seismic responses at the SS and FS of a stiffness eccentric structure based on experimentally validated structure hasn't yet been studied in detail. Therefore, influence of asymmetry on response variability at the FS and SS of the structure has been investigated and observations have been made from a statistical point of view. Since numerous researchers have explored that structural response due to stiffness eccentricity is of less significance in the inelastic range (Ferhi and Truman, 1996a), elastic seismic response of the structure has been considered under varying orientations of ground motion for a PGA of 0.4g.

Figure 7.1 shows a 3D model of the multi-story asymmetric structure considered for shaking table test. Point O is the origin of the orthogonal axes set for all floor axes X and Y in the same direction. The plane XOY is the same at all floor levels. Each floor has three degrees of freedom u_x, u_y and u_θ. The structure's equation of motion (Chopra, 2001) can be expressed as:

$$\text{M}\ddot{u}(t) + \text{C}\dot{u}(t) + \text{K}u(t) = \text{F}_{\text{eff}} \qquad (7.1)$$

where

$$\text{F}_{\text{eff}} = -\text{MI}\ddot{U}_g \qquad (7.2)$$

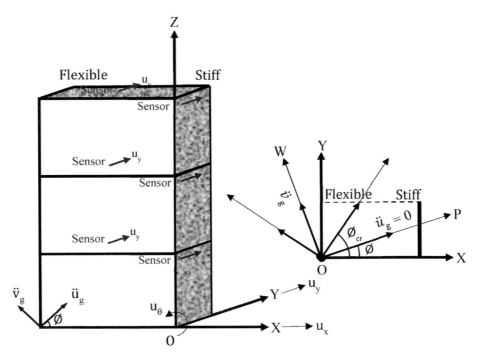

Figure 7.1 Structural 3D model and seismic orientations

M, C and K represent the global mass, damping and stiffness matrices of dimensions $3\psi'$ where ψ' is the degree of freedom. Each floor has been considered to have three degrees of freedom. If the DOF for the displacement vectors u_x, u_y and u_θ are ψ'_x, ψ'_y and ψ'_θ, then assumed degrees of freedom for structural floor can be expressed as:

$$\psi = \psi_x + \psi_y + \psi_\theta \tag{7.3}$$

Therefore, the overall displacement of the structure for the assumed DOF can be expressed as:

$$[u] = \left[\{u_x\}, \{u_y\}, \{u_\theta\}\right]^T \tag{7.4}$$

Assuming simplifications for each lateral resisting component of the structure, the structure's equation of motion can be developed by considering the stiffness matrix [K] as the sum of stiffness [∑K] of all resisting components relative to the overall coordinate system. Stiffness matrices for each lateral resisting component of the structure were developed with assumptions in the local degrees of freedom. The shear wall was assumed to have translational and rotational degree of freedom about a shear plane while the frame elements were assumed to have translational and rotational degrees of freedom about the vertical axis of the plane frame. Overall, based on these assumptions, it can be said that structural columns have two translational and one rotational degree-of-freedom about the orthogonal axes as illustrated in Figure 7.1. For determining the structural element's stiffness matrix of each lateral force component under local coordinates, the assumptions were made that shear wall has bending and shear deformation while the frame elements have only bending deformation.

The static condensation of the internal degrees of freedom was implemented to obtain the resisting force vectors which is an algebraic method (Bathe, 2006) incorporating a partial Gauss elimination. For reduction in the size of the equilibrium equations of a system, this method is used in finite elements with internal degrees of freedom. Hence, the degrees of rotation of the lateral force components were eliminated using static condensation method. After the application of static condensation, the ith stiffness matrix of the shear wall became a diagonal matrix with a dimension equal to 3ψ, where ψ represents the degree of freedom and the resulted relationship can be expressed as:

$$[\delta]_i = \begin{bmatrix} \delta_{li} \\ \delta_{ti} \\ \delta_{\theta i} \end{bmatrix} = \begin{bmatrix} \{k_{li}\} & 0 & 0 \\ 0 & \{k_{ti}\} & 0 \\ 0 & 0 & \{k_{\theta i}\} \end{bmatrix} \begin{bmatrix} \Delta_{li} \\ \Delta_{ti} \\ \Delta_{\theta i} \end{bmatrix} = [k]_i [\Delta]_i \tag{7.5}$$

where k_{li} and k_{ti} are the directional lateral stiffness matrices of the main orthogonal planes in the local coordinates, $k_{\theta i}$ is the torsional stiffness matrix in the local coordinates. δ_{li}, δ_{ti}, $\delta_{\theta i}$, Δ_{li}, Δ_{ti} and $\Delta_{\theta i}$ are the shear and displacement vectors corresponding to the longitudinal, transverse and vertical directions, respectively. After determining structural displacement vector into the global coordinates using transformation matrix, the structure's stiffness matrix in global coordinates can be expressed as:

$$[K] = \begin{bmatrix} [K]_x & [K]_{xy} & [K]_{x\theta} \\ [K]_{yx} & [K]_y & [K]_{y\theta} \\ [K]_{\theta x} & [K]_{\theta y} & [K]_\theta \end{bmatrix} \tag{7.6}$$

The undamped general equation of vibration using the developed stiffness matrix (Li *et al.*, 2004) is expressed in Equation 7.7 for two translational ground motions $\ddot{u}_g$ and $\ddot{v}_g$, and one vertical ground motion component $\ddot{\varphi}_{gz}$:

$$\begin{bmatrix} [m] & [0] & -[m][\alpha_y] \\ [0] & [m] & [m][\alpha_x] \\ -[m][\alpha_y] & [m][\alpha_x] & [J_0] \end{bmatrix} \begin{bmatrix} [\ddot{u}_x] \\ [\ddot{u}_y] \\ [\ddot{u}_\theta] \end{bmatrix} + \begin{bmatrix} [K]_x & [K]_{xy} & [K]_{x\theta} \\ [K]_{yx} & [K]_y & [K]_{y\theta} \\ [K]_{\theta x} & [K]_{\theta y} & [K]_\theta \end{bmatrix} \begin{bmatrix} [u_x] \\ [u_y] \\ [u_\theta] \end{bmatrix} =$$

$$- \begin{bmatrix} [m] & [0] & -[m][\alpha_y] \\ [0] & [m] & [m][\alpha_x] \\ -[m][\alpha_y] & [m][\alpha_x] & [J_0] \end{bmatrix} \begin{bmatrix} [\ddot{u}_g] \\ [\ddot{v}_g] \\ [\ddot{\varphi}_{gz}] \end{bmatrix} \tag{7.7}$$

where $[\alpha_x]$ and $[\alpha_y]$ are the global coordinates matrices at the centroid of the structural floor, and J_0 for centroid O of the floor is the moment of inertia matrix. It's a diagonal matrix, and its diagonal components can be expressed as:

$$J_{0i} = m_i \left(\gamma_i^2 + \alpha_{xmi}^2 + \alpha_{ymi}^2 \right) \tag{7.8}$$

where γ_i is the inertia radius at centroid α_{xmi} and α_{ymi} of the ith floor relative to the reference axes of the structure. Equation 7.8, given the coordinates of origin at any point, provides a coupled equation of vibration of multi-story structural system against the excitation of two-way horizontal and one rotational ground motion simultaneously acting on the structure. When the multi-story structure is used as the reference point, Equation 7.8 becomes:

$$\begin{bmatrix} [m] & [0] & -[m][\alpha_{ym}] \\ [0] & [m] & [m][\alpha_{xm}] \\ -[m][\alpha_{ym}] & [m][\alpha_{xm}] & [J_{cm}] \end{bmatrix} \begin{bmatrix} [\ddot{u}_x] \\ [\ddot{u}_y] \\ [\ddot{u}_\theta] \end{bmatrix} + \begin{bmatrix} [K]_x & [K]_{xy} & [K]_{x\theta} \\ [K]_{yx} & [K]_y & [K]_{y\theta} \\ [K]_{\theta x} & [K]_{\theta y} & [K]_\theta \end{bmatrix} \begin{bmatrix} [u_x] \\ [u_y] \\ [u_\theta] \end{bmatrix} =$$

$$- \begin{bmatrix} [m] & [0] & -[m][\alpha_{ym}] \\ [0] & [m] & [m][\alpha_{xm}] \\ -[m][\alpha_{ym}] & [m][\alpha_{xm}] & [J_{cm}] \end{bmatrix} \begin{bmatrix} [\ddot{u}_g] \\ [\ddot{v}_g] \\ [\ddot{\varphi}_{gz}] \end{bmatrix} \tag{7.9}$$

where $[J_{cm}]$ is the floor's moment of inertia with respect to the center of mass and $[\alpha_{xm}]$ and $[\alpha_{ym}]$ are the center of mass coordinate matrix relative to the seismic input points as the reference coordinates. As for the general multi-story structure, the centroid of the floors cannot be uniquely determined. Therefore, the coordinates of the origin at the centroid of the vibration equation is of no practical significance. For the damping matrix of the structure so far, the researchers have proposed several methods to determine the structural damping matrix (Chopra, 2001). Damping matrix for this research was expressed as:

$$[C] = a[M] + b[K] \tag{7.10}$$

$$[a] = \left(2(\xi_k \omega_l - \xi_l \omega_k)/\left(\omega_l^2 - \omega_k^2\right)\right)\omega_l \omega_k \tag{7.11}$$

$$[b] = 2(\xi_l \omega_l - \xi_k \omega_k)/\left(\omega_l^2 - \omega_k^2\right) \tag{7.12}$$

where ω_k and ω_l, respectively, indicate the first and third vibration frequency of the structure, and ξ_k and ξ_l respectively represent the first and third vibration damping ratio of the structure, i.e. k = 1 and l = 3.

By decomposing the force into its components, it can be expressed as:

$$F_{eff} = -M([I_P \ddot{u} + I_W \ddot{v}_g] + I_Z \ddot{\varphi}_{gz}) \tag{7.13}$$

where:

$$\begin{bmatrix} I_P \\ I_W \\ 0 \end{bmatrix} = \begin{bmatrix} \cos\theta & \sin\theta & 0 \\ -\sin\theta & \cos\theta & 0 \\ 0 & 0 & 1 \end{bmatrix} \begin{bmatrix} I_X \\ I_Y \\ I_Z \end{bmatrix} \tag{7.14}$$

Introducing Equation 7.14 into Equation 7.13, it can be re-written as:

$$F_{eff} = -\begin{bmatrix} [m] & [0] & -[m][\alpha_{ym}] \\ [0] & [m] & [m][\alpha_{xm}] \\ -[m][\alpha_{ym}] & [m][\alpha_{xm}] & [J_{cm}] \end{bmatrix} \begin{bmatrix} I_X\cos\theta + I_Y\sin\theta \\ -I_X\sin\theta + I_Y\cos\theta \\ I_Z \end{bmatrix} \begin{bmatrix} \ddot{u}_g \\ \ddot{v}_g \\ \ddot{\varphi}_{gz} \end{bmatrix} \tag{7.15}$$

Expanding Equation 7.15, the effective force can be expressed as:

$$F_{eff} = -\left\Vert \begin{bmatrix} [m] & [0] & -[m][\alpha_{ym}] \\ [0] & [m] & [m][\alpha_{xm}] \\ -[m][\alpha_{ym}] & [m][\alpha_{xm}] & [J_{cm}] \end{bmatrix} \begin{bmatrix} I_X\ddot{u}_g \\ 0 \\ 0 \end{bmatrix} \cos\theta + \right.$$
$$\begin{bmatrix} [m] & [0] & -[m][\alpha_{ym}] \\ [0] & [m] & [m][\alpha_{xm}] \\ -[m][\alpha_{ym}] & [m][\alpha_{xm}] & [J_{cm}] \end{bmatrix} \begin{bmatrix} 0 \\ I_Y\ddot{u}_g \\ 0 \end{bmatrix} \sin\theta +$$
$$\begin{bmatrix} [m] & [0] & -[m][\alpha_{ym}] \\ [0] & [m] & [m][\alpha_{xm}] \\ -[m][\alpha_{ym}] & [m][\alpha_{xm}] & [J_{cm}] \end{bmatrix} \begin{bmatrix} -I_X\ddot{v}_g \\ 0 \\ 0 \end{bmatrix} \sin\theta +$$
$$\begin{bmatrix} [m] & [0] & -[m][\alpha_{ym}] \\ [0] & [m] & [m][\alpha_{xm}] \\ -[m][\alpha_{ym}] & [m][\alpha_{xm}] & [J_{cm}] \end{bmatrix} \begin{bmatrix} 0 \\ I_Y\ddot{v}_g \\ 0 \end{bmatrix} \cos\theta +$$
$$\left. \begin{bmatrix} [m] & [0] & -[m][\alpha_{ym}] \\ [0] & [m] & [m][\alpha_{xm}] \\ -[m][\alpha_{ym}] & [m][\alpha_{xm}] & [J_{cm}] \end{bmatrix} \begin{bmatrix} 0 \\ 0 \\ I_Z\ddot{\varphi}_{gz} \end{bmatrix} \right\Vert \tag{7.16}$$

The final undamped general equation of vibration is as follows:

$$
\begin{bmatrix}
[m] & [0] & -[m][\alpha_{ym}] \\
[0] & [m] & [m][\alpha_{xm}] \\
-[m][\alpha_{ym}] & [m][\alpha_{xm}] & [\Omega_{cm}]
\end{bmatrix}
\begin{bmatrix}
[\ddot{u}_x] \\
[\ddot{u}_y] \\
[\ddot{u}_\theta]
\end{bmatrix}
+
\begin{bmatrix}
[K]_x & [K]_{xy} & [K]_{x\theta} \\
[K]_{yx} & [K]_y & [K]_{y\theta} \\
[K]_{\theta x} & [K]_{\theta y} & [K]_\theta
\end{bmatrix}
\begin{bmatrix}
[u_x] \\
[u_y] \\
[u_\theta]
\end{bmatrix}
=
$$

$$
-\left(
\begin{bmatrix}
[m] & [0] & -[m][\alpha_{ym}] \\
[0] & [m] & [m][\alpha_{xm}] \\
-[m][\alpha_{ym}] & [m][\alpha_{xm}] & [\Omega_{cm}]
\end{bmatrix}
\left(
\begin{bmatrix} I_X\ddot{u}_g \\ 0 \\ 0 \end{bmatrix}
+
\begin{bmatrix} 0 \\ I_Y\ddot{v}_g \\ 0 \end{bmatrix}
\right)
\right)\cos\theta +
$$

$$
\left(
-\begin{bmatrix}
[m] & [0] & -[m][\alpha_{ym}] \\
[0] & [m] & [m][\alpha_{xm}] \\
-[m][\alpha_{ym}] & [m][\alpha_{xm}] & [\Omega_{cm}]
\end{bmatrix}
\left(
\begin{bmatrix} -I_X\ddot{v}_g \\ 0 \\ 0 \end{bmatrix}
+
\begin{bmatrix} 0 \\ I_Y\ddot{u}_g \\ 0 \end{bmatrix}
\right)
\right)\sin\theta -
$$

$$
\left(
\begin{bmatrix}
[m] & [0] & -[m][\alpha_{ym}] \\
[0] & [m] & [m][\alpha_{xm}] \\
-[m][\alpha_{ym}] & [m][\alpha_{xm}] & [\Omega_{cm}]
\end{bmatrix}
\begin{bmatrix} 0 \\ 0 \\ I_Z\ddot{\varphi}_{gz} \end{bmatrix}
\right)
\tag{7.17}
$$

where:

$$
F^{gx}_{\ eff} = -
\begin{bmatrix}
[m] & [0] & -[m][\alpha_{ym}] \\
[0] & [m] & [m][\alpha_{xm}] \\
-[m][\alpha_{ym}] & [m][\alpha_{xm}] & [\Omega_{cm}]
\end{bmatrix}
\left(
\begin{bmatrix} I_X\ddot{u}_g \\ 0 \\ 0 \end{bmatrix}
+
\begin{bmatrix} 0 \\ I_Y\ddot{v}_g \\ 0 \end{bmatrix}
\right)
\tag{7.18}
$$

$$
F^{gy}_{\ eff} = -
\begin{bmatrix}
[m] & [0] & -[m][\alpha_{ym}] \\
[0] & [m] & [m][\alpha_{xm}] \\
-[m][\alpha_{ym}] & [m][\alpha_{xm}] & [\Omega_{cm}]
\end{bmatrix}
\left(
\begin{bmatrix} -I_X\ddot{v}_g \\ 0 \\ 0 \end{bmatrix}
+
\begin{bmatrix} 0 \\ I_Y\ddot{u}_g \\ 0 \end{bmatrix}
\right)
\tag{7.19}
$$

$$
F^{gz}_{\ eff} = -
\begin{bmatrix}
[m] & [0] & -[m][\alpha_{ym}] \\
[0] & [m] & [m][\alpha_{xm}] \\
-[m][\alpha_{ym}] & [m][\alpha_{xm}] & [\Omega_{cm}]
\end{bmatrix}
\begin{bmatrix} 0 \\ 0 \\ I_Z\ddot{\varphi}_{gz} \end{bmatrix}
\tag{7.20}
$$

Therefore, the general equation of motion of the structure would become:

$$
M\ddot{u}(t) + C\dot{u}(t) + Ku(t) = (F^{gx}_{\ eff})\cos\theta + (F^{gy}_{\ eff})\sin\theta + F^{gz}_{\ eff}
\tag{7.21}
$$

For a response quantity, utilizing the principle of superposition, the response history for any arbitrary seismic orientation $\varnothing$ may be reflected as a linear combination of three response histories. The first response history corresponds to F^{gx}_{eff}, the second corresponds to F^{gy}_{eff} and the third corresponds to the vertical excitation F^{gz}_{eff} (Athanatopoulou, 2005):

$$\Theta = \cos\varnothing \left(\Theta_{,1X} + \Theta_{,2Y}\right) + \sin\varnothing \left(\Theta_{,1Y} - \Theta_{,2X}\right) + \Theta_{,Z} \tag{7.22}$$

The general solution regarding time-dependent response quantity Θ can be expressed as:

$$\Theta\left(\varnothing, t\right) = \Theta_{,X}\left(t\right)\cos\varnothing + \Theta_{,Y}\left(t\right)\sin\varnothing + \Theta_{,Z}\left(t\right) \tag{7.23}$$

where:

$$\Theta_{,X}\left(\varnothing, t\right) = \Theta_{,1X}\left(\varnothing, t\right) + \Theta_{,2Y}\left(\varnothing, t\right) \tag{7.24}$$

$$\Theta_{,Y}\left(\varnothing, t\right) = \Theta_{,1Y}\left(\varnothing, t\right) - \Theta_{,2X}\left(\varnothing, t\right) \tag{7.25}$$

For the case of seismic excitations with $\ddot{u}_g\left(t\right)$ considered along the P-axis and $\ddot{v}_g\left(t\right)$ along W-axis and a vertical component $\ddot{\varphi}_{gz}$ for the actual ground motions, the seismic response Θ can be determined as $\Theta_{,P}$ and the preceding equation would become:

$$\Theta_{,P}\left(\varnothing, t\right) = \Theta_{,X}\left(t\right)\cos\varnothing + \Theta_{,Y}\left(t\right)\sin\varnothing + \Theta_{,Z}\left(t\right) \tag{7.26}$$

For the transposed ground motion components, Equation 7.26 would become:

$$\Theta_{,W}\left(\varnothing, t\right) = -\Theta_{,X}\left(t\right)\sin\varnothing + \Theta_{,Y}\left(t\right)\cos\varnothing + \Theta_{,Z}\left(t\right) \tag{7.27}$$

In the earlier discussion, 3D excitations have been discussed without any restraining condition corresponding to the seismic excitation. However, for special circumstances with various constraints, only part of them needs to be investigated and simplifications can be made. For example, the vertical component of seismic excitation has been ignored by numerous researchers (Kalkan and Kwong, 2013; Rigato and Medina, 2007; López and Torres, 1997). In this case, there were several constraints because of the limited capacity of the shaking table equipment. The seismic excitations were further simplified, and unidirectional seismic component was considered along the transverse direction of the structure ($\ddot{v}_g(t)$). Therefore, in this particular case, it was considered necessary to carry out the study based on the formation of the seismic angle $\varnothing$ of the unidirectional excitation with reference axis of the structure as shown in Figure 7.1. In this research since the seismic component along the direction of eccentricity is ignored ($\ddot{u}_g(t) = 0$), therefore, in Equations 7.24 and 7.25, the seismic responses $\Theta_{,1X}(t)$ and $\Theta_{,1Y}(t)$ can be reduced (Song et al., 2007). Thus, the typical response quantity in this particular case would become:

$$\Theta_{,X}\left(\varnothing, t\right) = \Theta_{,1X}\left(\varnothing, t\right) \tag{7.28}$$

$$\Theta_{,Y}\left(\varnothing, t\right) = \Theta_{,1Y}\left(\varnothing, t\right) \tag{7.29}$$

7.4   Validation of numerical model with experimental and theoretical results

There are two main reasons for the torsional vibrations caused by the earthquakes: one is the external disturbance; the movement of the seismic wave through the ground is extremely complicated and the period and phase of each point is different. Due to the difference in the movement between the ground particles, each part of the ground not only can produce a translational component but also produce a rotating component which forces the structure to produce torsional vibrations which are external. However, due to technical reasons, there is no real earthquake record of the ground motion component, so that the structural reversal effect caused by the previous reason is difficult to determine. Second, the building structure itself in the general structure of the seismic analysis is simplified into a flat model in its two main axis directions. It should be noted that even if the mass centers of the structural floors themselves coincide with the stiffness centers, and if they are not on a vertical line, the structure will produce torsional vibrations once exposed to translational excitations. Therefore, the preceding discussion concludes the fact that the nature of the seismic excitation is multi-dimensional and the building structure under the action of the seismic excitation is a complex problem of spatial vibration. The utilized experimental model is an eccentric frame-shear wall reinforced concrete structure. Due to the eccentricity in the structure, the C_M and C_R do not coincide at one single point. This eventually led to the structural damage during the shaking table test because of enhanced torsional response. The theoretical procedure ignored the out-of-plane stiffness of the shear wall. However, since the shear walls of many frame-shear wall structures are associated with the columns, it is actually equivalent to ignoring the out-of-plane stiffness of the frame columns connected to the shear walls. This can result in a smaller torsional rigidity of the theoretically calculated structure. Considering the out-of-plane stiffness of the frame columns, the lateral displacement and torsional rigidity of the shear wall was simulated by the out-of-plane stiffness of the two side edges of the structure.

The displacement history curves have been used to validate numerical results with theoretical and experimental results. In terms of the validation of the numerical model, global displacement response was prioritized over global acceleration demands. This priority was set because global acceleration demands were highly influenced by the structural cracking, stiffness degradation and non-uniform yielding. Compared with the acceleration demands, displacement demands were not influenced by local oscillations. In the case of acceleration demands, low-frequency global vibrations induced through the input excitation dominated the main waveform and caused the local oscillations to produce additional acceleration of high frequency with local effects. A nonlinear model of the frame-shear wall structure was simulated in SAP2000 (Wilson and Habibullah, 1998) and a moment resisting frame of experimental model was established according to the material properties, geometric properties and cross-sections of the frame components. In this chapter, a simplified nonlinear model of the shear wall structure was established. Therefore, the nonlinearity in the structure was established in the form of concentrated plastic hinges at member ends (Inel and Ozmen, 2006; Ladjinovic et al., 2012). The concentrated plastic hinges were based on moment-rotation (M-θ) bending and torsion degree of freedom relationship. Only frame members in the structure were expected to behave nonlinearly as from the experimental predictions the recorded strains in the shear wall were too low to cause any damage. Given that the primary goal of this chapter is the conceptual comparison of elasto-plastic time history responses with experimental and numerical responses, the nonlinear simplified model is appropriate and overall conclusions of the study aren't affected. The assumption of rigid floor diaphragm was incorporated at each story level. Structural components (i.e, columns and beams) were modeled implying the plastic hinges near beam-column joints. The structural components between the plastic hinges were assumed to be elastic (Fig. 7.2a). The behavior of the plastic hinge was implemented through moment-rotation curve (Lignos and Krawinkler, 2010),

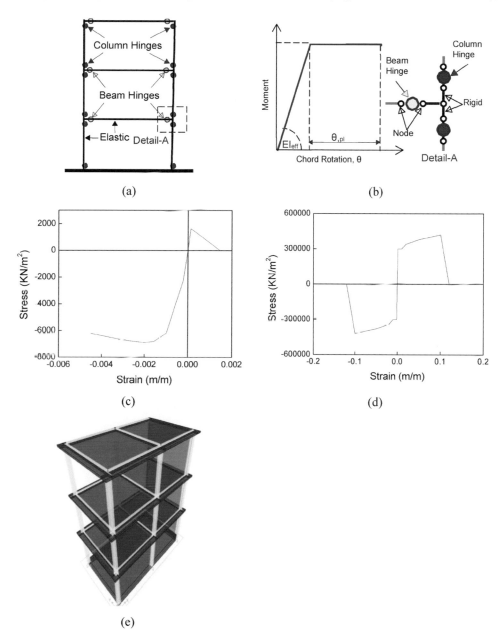

(a)

(b)

(c)

(d)

(e)

Figure 7.2 Analytical model for global response evaluation: (a) transverse frame with plastic hinges, (b) plastic hinge detail, (c) stress-strain relationship of concrete material, (d) stress-strain relationship of reinforcement and (e) FE model in SAP2000

with little simplification. According to the flexural capacities of columns and beams and their backbone relationships (Fig. 7.3b), modeling parameters were adopted. The elastic portions of the columns and beams were modeled with stiffness modifiers. The stress-strain relationship of the reinforcement bars (Park, 1985) was adopted in correspondence to the properties of HPB200 reinforcing steel using a strain hardening ratio of 0.01 and yield strength of 200 MPa (Fig. 7.3d). The stress-strain properties of the concrete were established using (Mander *et al.*, 1988) stress-strain model (Fig. 7.3c). For a better comparison of the numerical model, bending stiffness of the slabs and wall was implemented using shell elements. With

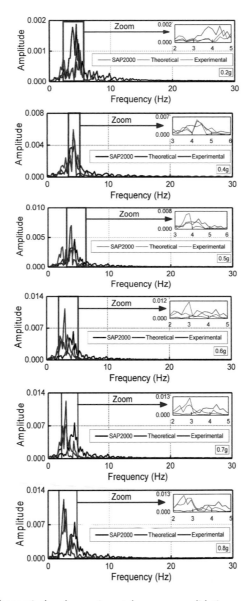

Figure 7.3 Numerical, theoretical and experimental response validation

an assumption of the neutral axis translation, the flexural stiffness was 20% of the gross sectional property of the slab (shell elements) and 15% of the total flexural stiffness was incorporated for the shear wall, and hence cracked sections were considered for flexural stiffness of the slabs and out-of-plane bending stiffness of the wall. Rayleigh damping of 5% was assumed in all modes. Key points that were considered in the modeling are as follows:

1   Beams and columns were modeled from centerline to centerline. Inelasticity was lumped as concentrated hinges at member ends followed by the bilinear moment-rotation envelopes.
2   The stiffness, strength and material behavior of the structural elements were modeled similar in the orthogonal directions.
3   Joints were considered as rigid.
4   The slab was considered as rigid diaphragm.
5   The connections of the columns with the base were considered as fixed.
6   Nonlinearities in the geometry were considered in the form of P-Δ effects in the numerical model.
7   Floor masses were lumped to all nodes with equal proportion.
8   5% damping was considered for all natural periods of the structure.

The numerical structure was exposed to the same El Centro 1940 earthquake record and the earthquake was progressively increased to obtain the desired displacement-history responses. The achieved displacement history results have been illustrated in the form of sum square amplitude. Numerical findings show a fairly good agreement with the findings of the test and theoretical results and have been illustrated in Figure 7.3. In spite of the good agreement of the investigated findings, there is some amount of errors at some peak shifts. Illustrations of Figure 7.3 report that numerical results are slightly higher than the experimental and theoretical results for PGA = 0.2g. Furthermore, Figure 7.3 also illustrates that numerical responses are marginally lower than the experimental and theoretical responses for PGA = 0.5g–0.8g.

Moreover, it can be seen that with the increase of the PGA, the peak frequency shift has moved slightly forward mainly because of the fact that the structure after a PGA of 0.6g entered into a highly inelastic state. The simplified nonlinear simulated model has not truly captured the unstable dynamic state of the structure after PGA of 0.6g despite the fact that the numerical findings are still in reasonably fair agreement with theoretical and experimental results. The reason behind this error is the simplified simulated model. Also, the material characteristics of the actual constructed structure may differ a bit with the material strengths due to poor construction practice. The other reasons have been explained in the next section.

7.5   Errors in theoretical and simulated results

The numerical model established in SAP2000 is based on the lumped plasticity model where only flexure and P-M interaction was considered to predict the overall nonlinear response of structure. The results obtained from SAP2000 moderately agreed with experimental results and deviated severely at some peak shifts mainly because of the assumptions considered in the modeling process. Since the aim of establishing the model in SAP2000 was to meet the global deformation criteria, local damage considerations were ignored and therefore, predicted response is not representative of the local damage conditions. The reason behind this is the complication associated with the non-uniform yielding behavior of the concrete structures. In such cases, accurate estimation of the effective EI and rotational ductility capacity in the orthogonal direction is highly complicated. The reason why the deformation demands moderately agreed with the experimental findings is the negligible influence of the local oscillations on the deformation demands compared with the acceleration demands where the influence of local oscillation is highly dominant. Moreover, for moment resisting frame elements, SAP2000 is a deformation-based software and therefore, implementing the actual internal forces at the damage location is not possible which leads to the differences in the experimental and numerical seismic response. This is clearly evident in Figure 7.3 that as the structure entered into a plastic state, the predicted deformation behavior started deviating from the experimental response. With further increase in the input excitation, the deviation further increased. Besides, in the actual structure, there always exist some weak parts, where the smaller forces cause internal cracking. The numerical and theoretical response considers the larger force of the crack and not the actual existence of the weak parts except the plastic hinges location in the numerical model. This yields in the difference between the actual situation and theoretical assumptions in the cracking time and place. In addition, the sliding between the model beam and the vibration table during the experiment also brings an error to the experiment, which is not reflected in the numerical and theoretical response. The deformation of the beam-column joint has not been taken as a point of interest in the numerical and theoretical results, which also had an impact on the results.

7.6   Response under varying orientations

Response variability depends upon both the seismic excitation and the response quantity. This fact has been illustrated in Figure 7.4 in which the FS and SS-wise distribution of response quantities over all possible seismic directions has been presented for the validated numerical model with PGA = 0.4g as shown in Figure 7.3. The considered quantities are maximum and minimum displacements, rotational displacements, velocities, rotational velocities, accelerations and rotational accelerations respectively, at top roof of the structure. The considered seismic action for all possible seismic directions is the same unidirectional seismic action that was being considered for the experiment. The seismic direction that was considered along the transverse direction of the structure to induce excitation during the experimental testing is termed as exp. orientation so that the seismic response from other orientations could easily be compared with the exp. orientation. The seismic responses in Figure 7.4 have been presented with respect to the center of mass of the structure where seismic response variability under varying orientations of seismic excitation can clearly be seen. Response quantities have been presented both in the form of translational response and rotational response in the local axes of the FS and SS frames of the structure. It can be seen that rotational responses yielded more variability than the translational responses which also confirms the fact that rotational responses are more influenced by the asymmetry and are the accurate representation of irregularity in the structure (Zhang *et al.*, 2016). Similarly, comparing the seismic response variability within the FS and SS frames, it can be seen that relative rotational response variation at the SS is higher than the relative rotational response variation at the FS of the structure. The computations were limited to a reasonable size, and the analyses were run from 0° to 360° with an increment of 10°. The ground motion was rotated anticlockwise. Also, the structural response when the seismic excitation was applied along exp. orientation was compared with the responses obtained from other angles. As a

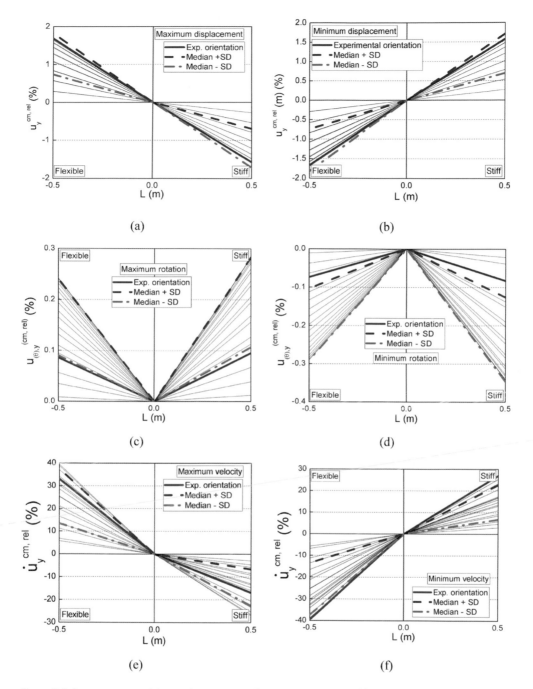

Figure 7.4 Response quantities under varying of seismic excitation: (a) maximum displacement, (b) minimum displacement, (c) maximum rotational displacement, (d) minimum rotational displacement, (e) maximum velocity, (f) minimum velocity, (g) maximum rotational velocity, (h) minimum rotational velocity, (i) maximum acceleration, (j) minimum acceleration, (k) maximum rotational acceleration and (l) minimum rotational acceleration; the thick, solid line corresponds to the response obtained when the seismic excitation was considered along exp. orientation; the dashed line corresponds to median + standard deviation response; the center line corresponds to the median − standard deviation response; the black lines correspond to other possible orientations

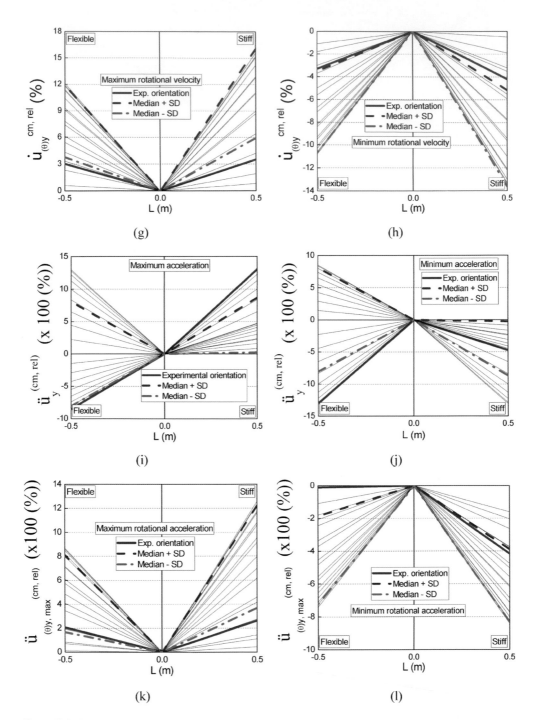

Figure 7.4 (Continued)

result of the exp. orientation, the response was compared with 18 different responses in the form of 21 FS- and SS-wise distributions of seismic response quantities. The distribution of seismic response quantities corresponding to exp. orientation has been highlighted with a thick, solid line.

It is quite clear that for almost all the response quantities the response obtained from exp. orientation has not led to the maximum response except for the maximum and minimum displacement response at the FS and SS edge of the structure where maximum seismic response occurred due to seismic excitation along exp. orientation. But for the same excitation, the structure's rotational displacement response appears to be the least of the responses out of all possible responses. Hence, it can be concluded that analyzing such kind of asymmetric structures along exp. orientation may lead to maximum translational response but may not lead to conservative rotational response. Similar observations of conservative response from exp. orientation were noticed for minimum velocity and maximum acceleration response. However, for the same response quantities, a non-conservative response was obtained for maximum velocity and minimum acceleration response. For rotational responses, it can be seen that seismic response under exp. orientation hasn't led to conservative response for all maximum and minimum response quantities. One interesting fact noted about the structure was that difference of variation in the rotational responses is very high compared with the variation difference at the FS edge of the structure. Looking for the same fact for translational responses, it can be seen that an utterly opposite trend was noticed with varying seismic orientations. Response variability was higher at the FS compared to the response variability at the SS of the structure. These results presented in Figure 7.4 endorse the fact that critical angle changes with the seismic excitation and response quantities. The reason behind this is the fact that critical angle is a kind of quantity which depends upon the complete response history of the response quantity. Therefore, it can be concluded that:

1   Determination of critical angle which would yield conservative structural responses simultaneously for both seismic excitation and response quantity is difficult. However, it is easy to estimate critical angle for a single response quantity under single ground motion pair and it is explained in upcoming sections.
2   Different earthquakes excite different modes. Therefore, the response of a structure depends upon the characteristics of the earthquake and modal properties of the structure.
3   Seismic excitation along reference axis (exp. orientation) of the structure may lead to conservative seismic response but at the same time rotational response of the system may not provide a conservative response and the rotational response may even correspond to the least response over other possible seismic directions.

7.7   Is there any need to consider various orientations of seismic excitation?

To address this issue, the seismic excitation along the exp. orientation was considered along all other possible directions, and the seismic response was evaluated from a statistical viewpoint. Therefore, response quantities have been illustrated as box and whisker plots in Figure 7.5 for better investigation of the sample obtained through various excitations. The response quantities have been enclosed between 1.5 times the inter-quartile-range (IQR). The response quantities have been enclosed as a box and whisker plot so as to have a look

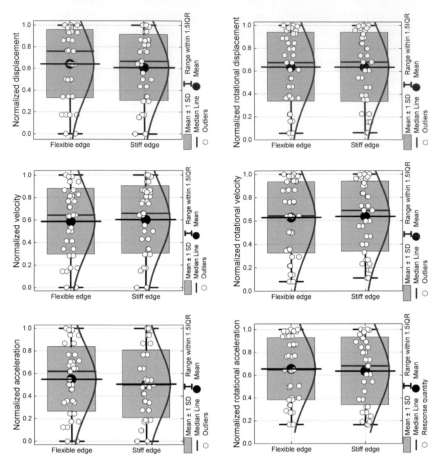

Figure 7.5 Box and whisker plot of response quantities at top roof level of the structure; the thick, solid line corresponds to the normal distribution of the response quantities

at how spread out the response quantities are. The variability of the response quantities has been demonstrated in the form of outliers while the skewness of the quantities has also been represented in the form of normal distribution (thick, solid line). The median response for most of the rotational response quantities could be found higher than the other response quantities which indicates the existence of higher response quantities for most of the orientations. It can also be noted that the median response for rotational response quantities at the SS is higher than the FS response quantity except the displacement response where the median response was found to be equal at both the edges. For other response quantities, it has been found that median response at both the edges of the structure has a mixed response. For example, for displacement and acceleration response quantities, the median response at the FS is higher compared with the SS while for the velocity response, both the edges demonstrated a similar response trend. This also indicates that rotational response quantities have more variation under varying orientations of ground motion compared with translational response quantities.

For better understanding of variation in the response quantities, the responses have been shown in the form of histograms. Assume for instance that varying orientation of the seismic excitation is the only cause of uncertainty of the seismic responses. In other words, for a structure under excitation, the probability distribution of the response quantity will be directly related to the varying orientation of seismic excitation. To evaluate the usefulness of varying orientations of seismic excitation along other directions, this conditional distribution for the response quantity has been used to serve as a benchmark.

Because for each response quantity of interest the functional relationship is different between varying orientation of seismic excitation and the response quantity, various response quantities will have different probability distributions. Furthermore, this relationship is more complicated in general, especially for nonlinear inelastic structures. Therefore, direct computation of the probability distributions is not possible. A random sample of orientations is generated assuming the varying orientation as a uniformly distributed random variable. The response quantity of interest is determined for the random sample of orientations. Figure 7.6 illustrates the summary of such data in the form of histograms, for all possible directions

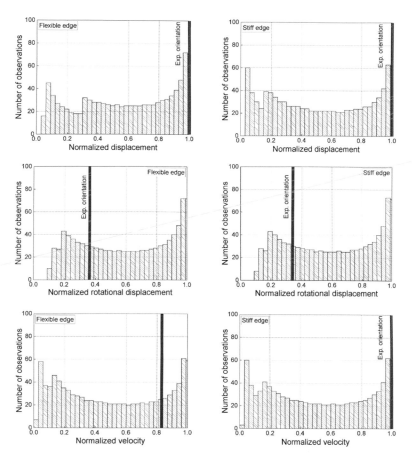

Figure 7.6 Histograms of 1000 randomly obtained realization of response quantities at top roof level of the structure; the thick, solid line corresponds to the response quantity obtained through exp. orientation

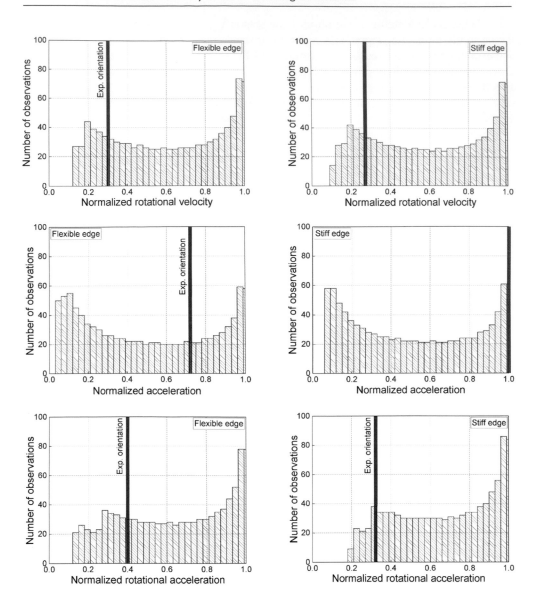

Figure 7.6 (Continued)

and for all considered response quantities at top roof level of the structure. The histograms are the representation of approximate probability distributions of the normalized response quantities. The response quantities were normalized with maximum value. The dependence of response variability on response quantity of interest and orientation of seismic excitation is confirmed by the normalized scale. These approximate histograms include a possible range of finite orientations of seismic excitation. Most of the approximate histograms share a common shape for translational responses. Similar observations were made for rotational responses.

7.8 Statistical distribution of structural response under varied orientations

Uncertainty in the seismic response due to varying orientations of ground motion has also be quantified by Kalkan and Kwong (2013) and Magliulo *et al.* (2014). Particularly, the illustrations of Figure 7.7 show the percentiles of the response quantities based on the normal distribution with 95% confidence for upper and lower bounds. The defined distributions are

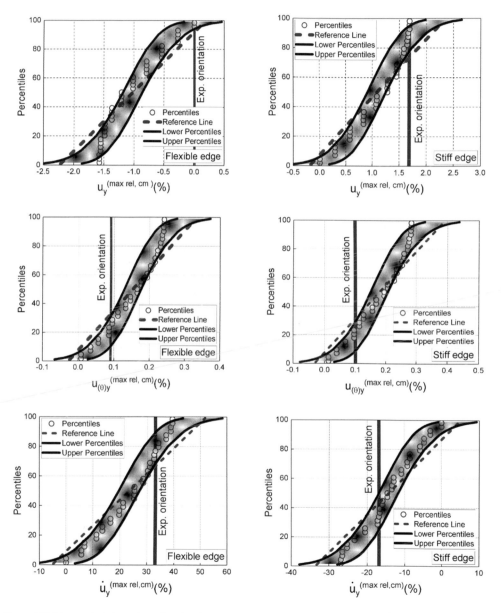

Figure 7.7 The probability of randomly observing response quantities at top roof level of the structure; the thick, solid line corresponds to the response quantity obtained through exp. orientation

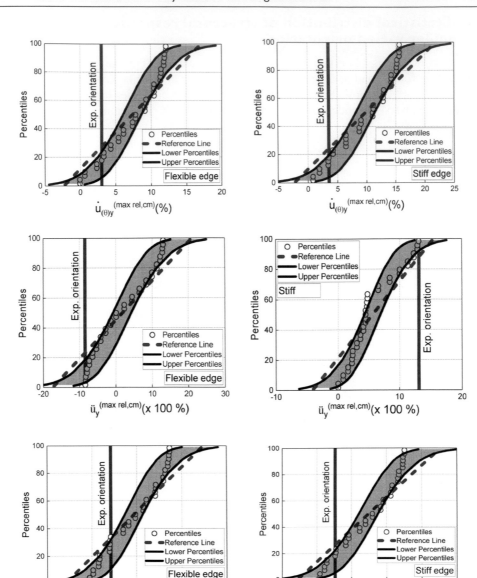

Figure 7.7 (Continued)

based on the median-value and standard deviation ρ of response quantities. The slope of the distribution curve is based on the standard deviation ρ which reflects the variation (uncertainty) in the response quantity of interest.

Assume for an instance that the only cause of response uncertainty is the unpredictability of the orientation of seismic excitation. Simplifying this assumption, it can be said that the probability distribution of the seismic orientation has a direct relation with the probability distribution of response quantity of interest. This hypothesis can be considered to assess the influence of variability in the orientation of seismic excitation as demonstrated by Kalkan and Kwong (2013) and Magliulo *et al.* (2014). Because the functional relationship between the response quantity and seismic orientation is different for each response quantity, various probability distributions would be obtained for multiple response quantities.

To quantify this observation, the idea of evaluating the seismic response in terms of percentiles was explored. For the relative response quantities, approximate percentiles have been shown in Figure 7.7. The thick, solid line indicates the response obtained when the excitation was applied at exp. orientation. To understand what information this line illustrates, we demonstrate the very first subplot in Figure 7.7. In the mentioned subplot, there lies approximately 20% probability of observing top roof displacement response when the seismic excitation was considered along the exp. orientation. Eventually, this describes the fact that there is underestimation of displacement response with approximately 80% probability. Now observing all the response quantities along exp. orientation, it is clear that the likelihood of finding the response quantities higher than the response quantities of exp. orientation is consistently higher than 80% for most of the response quantities. However, this trend does not hold true for displacement response at the SS, velocity response at the FS and acceleration response at the SS. For rotational responses, it can be seen that none of the subplot yielded maximum probability for exp. orientation.

7.9   Summary

In this chapter, the influence of varying orientations of ground motion using a validated numerical model was carried out on a mono-eccentric TS structure. The experimental results were utilized to validate the numerical and theoretical results. Notably, the numerical and theoretical responses were aimed at developing the precise prediction of the displacement responses. The results showed that both numerical and theoretical analyses predicted the displacement responses of experimental structure reasonably well when compared with experimental results. However, both the numerical and theoretical responses were based on some limitations as described in the preceding discussion and the responses deviated from the experimental results at some peak shifts of the displacement response. A study of response variability based on the varying orientations of seismic excitation was developed using the numerical model at PGA of 0.4g which led to the following conclusions:

- Comparing the relative displacement, velocity and acceleration responses, it was observed that the mono-eccentric frame-shear wall structure appears to have more sensitivity towards acceleration response than the velocity and displacement response. The sensitivity of the acceleration response was observed to be higher at the FS than the SS. Comparing the same response quantities in terms of response variability, acceleration response has a higher response variability than other response quantities. However, the acceleration variability greatly influenced the response at the FS with respect to the SS of the structure.
- Comparing the rotational displacement, rotational velocity and rotational acceleration responses, it was observed that the mono-eccentric frame-shear wall structure appears to have more sensitivity towards all the rotational response quantities. For rotational response quantities, the SS of the structure appears to be more sensitive than the FS of the structure. Moreover, comparing the same response quantities in terms of response

variability, it was observed that the rotational response variation has higher influence on the SS of the structure with respect to the response variability at the FS.

- Experimental orientation (exp. orientation) of the seismic input led to the maximum response for displacement. However, the same excitation led to the least rotational displacement response which proves the high uncertainty of the response quantities even for the same orientation of the seismic excitation. It can be said that even though exp. orientation led to the maximum displacement response, it was simultaneously non-conservative for other response quantities especially the rotational response quantities.
- Rotational responses for all the rotational response quantities were found to have been close to minimum among other orientations of the seismic excitation. Besides this, the response variability for rotational response quantities influenced both the edges of the structure.
- Treating the exp. orientation as a randomly selected orientation, it was observed that there is approximately 80% probability that for most of the rotational response quantities, the seismic response will exceed the exp. orientation's response. Similar observations were noticed for displacement, velocity and acceleration response with a probability of exceedance of 60% approximately.

For most of the relative response quantities, the probability of exceedance appeared to be lower at the SS compared to the response at the FS of the structure.

Chapter 8

Seismic design guidelines for asymmetric structures

8.1 Introduction

The level of damage in irregular buildings investigated in previous studies suggests that more seismic design guidelines are required on the avoidance of failures. It is the need of the hour that such guidelines are presented as a positive part of the revision for corresponding torsional provisions in the current seismic design codes. Specifically, many of the buildings that collapsed in past earthquakes had both soft and weak floor irregularity. The code rules for buildings with these irregularities were prompted by observations of that damage, but explicit collapse assessments for a broad design space of buildings with multiple irregularities have not yet been performed. Moreover, it is still unknown as to what extent the code defined provisions are effective under the interaction of different irregularities.

This chapter focuses on the following objectives:

- Determining how effective are the quantitative triggers defined by various seismic design codes mainly including ASCE/SEI (2016) for the design requirements of irregular structures and protection of the existing building infrastructure from the possible seismic hazard.
- Assessing and proposing modifications, if any, to the quantitative triggers for structural irregularities to address the conservative and non-conservative torsional provisions.

For the ease of readers, in the tables corresponding to the assessment of torsional guidelines, ASCE/SEI (2016) has been mentioned as it can be seen in Chapter 2 that most of the irregularity triggers are relatively the same in the majority of the seismic design code. Since ASCE/SEI (2016) provides better representation and in-depth explanation of irregular structures, it has been quoted for comparison purposes. However, design-based suggestions are not limited to ASCE/SEI (2016) and can be referred to other seismic design codes as well.

In addition to the damage response assessment at both local and global levels along with the evaluation of existing torsional design provisions of irregular structures, the objective of this book is to propose some revisions for the circumstances that are not yet well addressed by the seismic design codes. Following are the specific focuses of this chapter:

- To eliminate unnecessary sources of high conservatism where possible for the torsional design of irregular structures. This includes relaxing the quantitative triggers for structural irregularities. Similarly, to address the circumstances where torsional provisions are non-conservative for the possible seismic safety of irregular structures. This includes several suggestions and revision in the quantitative trigger for structural irregularities.

- Suggestions for the method of analysis for the design of irregular structures. This includes the prohibition of equivalent lateral force (ELF) method of analysis for the various unexplored circumstances that exist in real-life structures. Moreover, suggestions for orthogonality influence evaluation have been proposed due to the possible coupling of seismic response in horizontal directions.
- As per GB50011-2001, structures with planar and/or vertical irregularities shall be treated as especially irregular structures. Such structures are required to be designed considering locally enhanced deformations. Considering the provisions stated in section 3.4.4 of GB50011-2001, much more efficient strengthening measures are required to be taken through special studies. The provision further requires the determination of the weak locations for performance-based design. The suggested provision can only be adopted once the damage characteristics at higher PGAs are known.

8.2   Contribution of this chapter to knowledge

This chapter is mainly inspired by the following questions:

- When are the torsional provisions too conservative for irregular structures? And when do they become non-conservative?
- How are these provisions effective in determining location-specific consequences of irregularities? And how can they help in finding out weak locations in the existing irregular structures? This question is very important corresponding to the fact that damage to irregular structures during an earthquake is linked with the damage initiation at the weak location of the irregular structure.
- Are certain prohibitions necessary for seismic design categories (SDC) A, B, C, D, E and F defined by ASCE/SEI (2016)?

8.3   Description of the utilized parameters of irregular structures

Since damage assessment is one of the primary focuses of this book, data collected from the extensive experimental investigation of various asymmetric structures is used to propose design guidelines in the context of expected damage in various categories of asymmetric structures. The presented design guidelines are attributed to each damage level observed in this book and hence the damage response observed at each PGA is contributory towards the proposed design and structural health monitoring guidelines. The proposed recommendations for the design of asymmetric structures are based on several parameters described in Tables 8.1 and 8.2.

Most of the previous research data stock is only based on each irregularity present separately in the structure because of a wide range of characteristics involved in the estimation of seismic response. Therefore, it should be noted that further research is required for irregular structures with their characteristics not falling into the described criteria used in this book. Moreover, for the inclusion of the equivalent static torsional forces in equivalent lateral force method (ELF), a detailed investigation is required. It is noteworthy to mention that numerous international building codes (ASCE/SEI 16, UBC, 1997) have already defined a method to include torsional effects for equivalent lateral static force procedure by means of shifting the center of mass of the structure to some percentage that is a multiple of the

Table 8.1 Characteristics variation in the experimental models

Characteristics	Description
Building height	Low rise (three floors)
Degree of irregularity	Low RI to highly RI and IRI
Analysis method used for the design of experimental models	Equivalent lateral force analysis
Moment resisting frame (MRF)	RC shear wall frame (RC Special-MRF) Steel frame (Ordinary MRF) Steel frame (Special MRF)
Type of torsional system	From highly TS to highly TF

Table 8.2 Monitored metrics in the experimental models

Metric	Description
Strain time histories and residual strains for all types of demonstrated irregularities	Experimentally evaluated at the FS and SS of the TU-TS and TU-TF structures
Dynamic characteristics of the damaged model	Correlation of frequency and damping ratio behavior with other global and local seismic demands
Partial collapse	Experimentally determined
	Experimentally simulated for different types of irregularities
	Numerical simulation for FS and SS behavior
The direction of seismic excitation	Numerical simulation along with statistical evaluation
Floor displacements	For C-1 model under 1/3MCE, 2/3MCE, MCE and above
Floor accelerations	For all models under 2/3MCE, MCE and above
Ductility demand	For C-1 model under 1/3MCE, 2/3MCE, MCE and above
FS and SS rotation	For all models under 2/3MCE, MCE and above

plan dimension. However, in the opinion of the author, such methods are ineffective under extreme seismic events due to the following reasons:

1   Such methods improve the global behavior of the structure only and do not address the local torsional effects.
2   Treating both the flexible and stiff edge of the asymmetric structures in a similar fashion. Extensive details have been given in this book, which advocates separate treatment for both the edges.
3   Floor eccentricities transmit their influence to the adjacent floors. In this regard, this method is incapable of considering such effects especially in terms of the local seismic response.

Figure 8.1 Seismic risk levels under torsional induced vibrations

| Top to bottom / bottom to top | No influence on the adjacent floors |

Figure 8.2 Eccentricity influence on the adjacent floors

Relative damage performance of various irregular structures is presented in this book and based on the research findings; the relative level of seismic hazard along with analysis and design-based recommendations is demonstrated in Tables 8.3 through 8.8. The terms KI_1^{s-1}, VI_2^{s-2}, MI_2^{s-1}, SI_3^{s-1}, WI_4^{s-1}, CI_5^{s-1} and BI_6^{s-1} used in these tables refer to the stiffness irregularity, mass irregularity, soft floor mechanism, weak floor mechanism, re-entrant corners and structure with converged C_M and C_R but dislocated from C_G respectively. The subscripts 1, 2, ... 6 before the parenthesis refer to the number of irregularity parameter while the superscripts refer to the type of structures mainly considered for these parameters. The subscripts after the parenthesis refer to the direction of irregularity, i.e. planar and vertical. Seismic hazard under torsional induced vibrations is represented through green and yellow colors in Figure 8.1.

Furthermore, for the purpose of seismic design guidelines, it is demonstrated through Figure 8.2 as to how irregularities transfer their influence on the adjacent floors. A green color mentioned for lower floor eccentricity indicates that the eccentricity of the lower has either transmitted its influence to the adjacent floor or it has caused the highest response at the top roof level. Similarly, yellow color for the same eccentricity indicates that only the floor containing eccentricity is affected by the torsional induced vibrations and the eccentricity hasn't transmitted its influence to the adjacent floors. On the other hand, for eccentricity on the top roof level, the same description is true except the direction of influence which in this case is reversed. For green color, it indicates either influence transfer from top to bottom or maximum response at the bottom floor level when the eccentricity is actually at the top roof level.

8.4   Design guidelines

Based on the findings of this research, the following has been concluded for TS-RI structures:

- The prohibition of extremely TU buildings in SDC E and F defined in ASCE/SEI (2016), as long as the lateral system is proportioned according to the recommendations, is too conservative. For these types of structures, irregularity is only justified in terms of the quantitative trigger. In terms of its influence on the seismic response, negligible enhancement in the response is observed.
- The prohibition of equivalent lateral force (ELF) procedure on the basis of torsional irregularity is too conservative.
- Code defined mass irregularity is too conservative for such kind of structures.
- Orthogonality influence is required to be considered as in some cases the two directional translational mode response appears to be coupled.

In the similar structure described in Table 8.3 when the irregularity profile was changed from RI to IRI, a slight enhancement in the seismic response was observed. This means that the interaction of irregularities has adverse effects on the seismic response of TS structures. However, since the structure is TS, the seismic response is not augmented to a great extent. In Table 8.4, detailed observation and design recommendations considering IRI profile of irregularity for TS structures are presented. Moreover, it should be noted that in this case, the seismic response is critical at the SS of the structure and not the FS of the structure. Additionally, the location-specific influence of irregularity is higher compared with over all global response of the structure.

In Table 8.5, most critical seismic structures in the category of TS structures are demonstrated. Seismic hazard in the particular case is relatively higher compared with previous categories. It is also evident that the interaction of vertical mass irregularity with soft and (or) weak floor mechanism leads to higher seismic hazard and indicates more chances of damage at the FS of the structure. Though local tensile deformations are higher at the SS as well but looking at the overall picture of the local tensile deformations and global behavior of the structure, the expected hazard level at the FS can be predicted. Moreover, it is suggested to take extra precautions for such structures where C_M and C_R are converged but dislocated from the C_G of the structure as the local and floor-specific seismic response is highly enhanced in such scenarios. Therefore, the following is recommended for the design process of such kind of structures:

- Code-based vertical mass irregularity should be considered $M_i < 1.3\ M_a$ when the weak/soft floor is present in the structure as the interaction of irregularities leads to highly enhanced response under seismic excitations.
- ELF method of analysis should be prohibited for the SDC D, E and F as the method is based only on the fundamental mode of the structure and cannot consider complex interaction of irregularities under seismic excitations.
- It is suggested to use critical seismic response based on seismic excitation along all possible directions of the lateral moment resisting systems for SDC D, E and F.

In Table 8.6, design recommendations for TS structures having re-entrant corners have been presented. A code defined measure of irregularity for re-entrant corners is way too conservative and the quantitative trigger based on the research findings is suggested to be $C_i \leq$ 30% as seismic hazard in terms of damage response in such kind of structures is negligible. Therefore, the following is recommended for the design process of such kind of structures:

- Code-based quantitative trigger for re-entrant corners irregularity is suggested to be $C_i \leq 30\%$.
- It is suggested to use critical seismic response based on seismic excitation along all possible directions of the lateral moment resisting systems for SDC E and F.

In Table 8.7, it is demonstrated that interaction re-entrant corner irregularity with other irregularities under IRI profile is a major trigger for seismic hazard even if the structure is TS. The following recommendations for the design process of such kind of structures are presented:

- Code-based quantitative trigger for re-entrant corners irregularity is suggested to be $C_i \leq 15\%$ and is justifiable in such kind of structures.
- It is suggested to use critical seismic response based on seismic excitation along all possible directions of the lateral moment resisting systems for SDC B, C, D, E and F.

In Table 8.8, seismic design recommendations have been proposed for the most critical scenario. TF structures with mass, stiffness and strength eccentricities and weak/soft floor

Table 8.3 TS structures with RI profile of irregularities

Type of irregularity	KI_1^{s-I}, C^{-I}_P	MI_2^{s-I}, C^{-I}_P	SI_3^{s-I}, V	WI_4^{s-I}, V	CI_5^{s-I}	BI_6^{s-I}, P
Quantitative/qualitative trigger	Displaced C_R from C_G	Displaced C_M from C_G	$K_i < 0.7K_{i+1}$	$K_i < 0.65K_{i+1}$	$C_i \leq 15\%$	Converged C_M and C_R but dislocated from C_G
	Global seismic risk level		N/A	N/A	N/A	
	Location-specific seismic risk level/FS		N/A	N/A	N/A	
	Location-specific seismic risk level/SS		N/A	N/A	N/A	
	Floor-specific seismic risk level (floor containing eccentricity)		N/A	N/A	N/A	N/A
	Irregularity influence at the adjacent floors when the irregularity is in the lower floor		N/A	N/A	N/A	N/A
	Irregularity influence at the adjacent floors when the irregularity is at the top roof level		N/A	N/A	N/A	N/A
Remarks	Code defined mass irregularity is way too conservative both in terms of local and global damage response for RI type TU-TS structures. If the vertical mass irregularity is solely present in the structure and is present only on one floor, the proposed quantitative trigger based on this research is recommended as $M_i < 1.75\,M_a$.					
Recommended analysis procedure						
3D analysis recommendation	All SDC defined by ASCE/SEI (2016)	All SDC defined by ASCE/SEI (2016)	N/A	N/A	N/A	–
Prohibition of equivalent lateral force method	–	–	N/A	N/A	N/A	–
Orthogonality influence	All SDC defined by ASCE/SEI (2016)	–	N/A	N/A	N/A	–

Left-margin labels: TS-RI irregular Structures with RI profile of irregularity · Seismic risk triggers for various irregularities

Linear/nonlinear dynamic analysis	—	—	N/A	N/A	N/A	—
Design recommendations						
Torsional irregularity based on rotation instead of drift demand	—	—	N/A	N/A	N/A	—
Consideration of structural irregularity based on location-specific influence	SDC B through F defined by ASCE/SEI (2016)	—	N/A	N/A	N/A	—
All possible directions of seismic excitation (critical response)	SDC D and E ASCE/SEI (2016)	—	N/A	N/A	N/A	—
Detailing			—			
Recommendations for the design of new irregular structures	Avoiding the presented irregularities for SDC type E structures					
Recommendations for the health monitoring of existing irregular structures	The contour plots presented in this research are proposed to be incorporated for expected weak locations in existing irregular structures					

Table 8.4 TS structures with IRI profile of irregularities

Type of irregularity	$(Kl_1^{s-1}, C-1 + Ml_2^{s-1})_P$	$(Kl_1^{(s-1,C-1)})_P + (Ml_2^{(s-1)})_V$	$Sl_3^{s-1} V$	$Wl_4^{s-1} V$	Cl_5^{s-1}	$Bl_6^{s-1} P$
Quantitative/Qualitative trigger	Displaced C_R and C_M from C_G	Displaced C_R from C_G and $M_i < 1.5\,M_a$	$K_i < 0.7K_{i+1}$	$K_i < 0.65K_{i+1}$	$C_i \leq 15\%$	Converged C_M and C_R but dislocated from C_G
Global seismic risk level			N/A	N/A	N/A	
Location-specific seismic risk level/FS				N/A	N/A	
Location-specific seismic risk level/SS			N/A	N/A	N/A	
Floor-specific seismic risk level (floor containing eccentricity)			N/A	N/A	N/A	
Irregularity influence at the adjacent floors when the irregularity is in the lower floor			N/A	N/A	N/A	N/A
Irregularity influence at the adjacent floors when the irregularity at the top roof level floor			N/A	N/A	N/A	N/A
Remarks	Code defined mass irregularity proves to be justified for TS structures having IRI profile. However, seismic hazard is still not too big despite the interaction of irregularities.					
Recommended analysis procedure						
3D analysis recommendation	All SDC defined by ASCE/SEI (2016)	All SDC defined by ASCE/SEI (2016)	N/A	N/A	N/A	All SDC defined by ASCE/SEI (2016)
Prohibition of equivalent lateral force method	–	–	N/A	N/A	N/A	–
Orthogonality influence	SDC C, D, E and F defined by ASCE/SEI (2016)	All SDC defined by ASCE/SEI (2016)	N/A	N/A	N/A	–

Row group label: TS-IRI structures — Seismic risk triggers for various irregularities

			SDC E and F defined by ASCE/SEI (2016)	N/A	N/A	N/A	
Linear/nonlinear dynamic analysis	–	–		N/A	N/A	N/A	–
Design recommendations							
Torsional irregularity based on rotation instead of drift demand	–	–		N/A	N/A	N/A	–
Consideration of structural irregularity based on location-specific influence	All SDC defined by ASCE/SEI (2016)	–		N/A	N/A	N/A	–
All possible directions of seismic excitation (critical response)	SDC D and E ASCE/SEI (2016)	–		N/A	N/A	N/A	–
Detailing	Collectors are recommended to be used for SDC type E structures to cater with excessive tensile deformations at the FS of the structure and to facilitate force transfer mechanism						
Recommendations for the design of new irregular structures	It is suggested not to prohibit ELF method of analysis for SDC E and F in such kind of structures as the influence of structural irregularity towards damage is not too big provided the structures are designed considering appropriate lateral force resisting system						
Recommendations for the health monitoring of existing irregular structures	The contour plots presented in this research are proposed to be incorporated for expected weak locations in existing irregular structures of such types						

Table 8.5 TS-IRI structures with soft/weak floor

Type of irregularity		$(KI_l^{s-1} + MI_2^{s-1})_P$	$(KI_l^{(s-1)})_P + (VI_2^{(s-1)})_P$	$SI_3^{s-1}\ \vee$	$WI_4^{s-1}\ \vee$	CI_5^{s-1}	$BI_6^{s-1}\ \vee\ P$
	Quantitative/qualitative trigger	Displaced C_R and C_M from C_G	Displaced C_R and C_V from C_G	$K_i < 0.7K_{i+1}$	$K_i < 0.65K_{i+1}$	$C_i \leq 15\%$	Converged C_M and C_R but dislocated from C_G
Seismic risk triggers for various irregularities		Global seismic risk level					
		Location-specific seismic risk level/FS				N/A	
		Location-specific seismic risk level/SS				N/A	N/A
		Floor-specific seismic risk level (floor containing eccentricity)				N/A	N/A
		Irregularity influence at the adjacent floors when the irregularity is in the lower floor				N/A	N/A
		Irregularity influence at the adjacent floors when the irregularity is at the top roof level floor				N/A	N/A
	Remarks	Code defined mass irregularity is found to be highly non-conservative in this case. The proposed quantitative trigger for vertical mass irregularity based on this research is recommended as $M_i < 1.3\ M_a$.					
		Recommended analysis procedure					
	3D analysis recommendation	All SDC	All SDC	All SDC	All SDC	N/A	All SDC
	Prohibition of equivalent lateral force method	D, E and F	SDC D, E and F	SDC D, E and F	SDC D, E and F	N/A	SDC E and F
	Orthogonality influence	All SDC	All SDC	All SDC	All SDC	N/A	All SDC
	Linear/nonlinear dynamic analysis	SDC D, E and F	SDC D, E and F	SDC D, E and F	SDC D, E and F	N/A	SDC E and F
		Design recommendations					
	Torsional irregularity based on rotation instead of drift demand	SDC E and F	SDC E and F	SDC E and F	SDC E and F	N/A	—

TS-IRI structures

Consideration of structural irregularity based on location-specific influence	All SDC	All SDC	All SDC	N/A	All SDC
All possible directions of seismic excitation (critical response)	SDC D, E and F	SDC D, E and F	SDC D, E and F	N/A	–
Detailing	Collectors are recommended to be used for SDC D, E and F structures to cater with excessive tensile deformations at the FS of the structure and to facilitate force transfer mechanism. Moreover, the FS frame of the structures is recommended to be braced.				
Recommendations for the design of new irregular structures	Strictly avoiding the presented irregularities for SDC type E and F				
Recommendations for the health monitoring of existing irregular structures	The contour plots presented in this research are proposed to be incorporated for expected weak locations in existing irregular structures				

Table 8.6 TS-RI structures with re-entrant corners

Type of irregularity	KI_1^{s-1} P	MI_2^{s-1} P	SI_3^{s-1} V	WI_4^{s-1} V	CI_5^{s-1}	BI_6^{s-1} P
Quantitative/qualitative trigger	Displaced C_R from C_G	Displaced C_M from C_G	$K_i < 0.7K_{i+1}$	$K_i < 0.65K_{i+1}$	$C_i \leq 15\%$	Converged C_M and C_R but dislocated from C_G
Global seismic risk level			N/A	N/A		
Location-specific seismic risk level/FS			N/A	N/A	N/A	
Location-specific seismic risk level/SS		N/A	N/A	N/A	N/A	
Floor-specific seismic risk level (floor containing eccentricity)			N/A	N/A	N/A	
Irregularity influence at the adjacent floors when the irregularity is in the lower floor			N/A	N/A	N/A	N/A
Irregularity influence at the adjacent floors when the irregularity is at the top roof level floor			N/A	N/A	N/A	N/A
Remarks				–		
3D analysis recommendation	All SDC	All SDC	N/A	N/A	All SDC	All SDC
Prohibition of equivalent lateral force method	–	–	N/A	N/A	N/A	–
Orthogonality influence	–	–	N/A	N/A	N/A	–
Linear/nonlinear dynamic analysis	–	–	N/A	N/A	N/A	–
Torsional irregularity based on rotation instead of drift demand	–	–	N/A	N/A	N/A	–

Row grouping labels: *Seismic risk triggers for various irregularities* (trigger and seismic risk level rows); *Recommended analysis procedure* (3D analysis through linear/nonlinear dynamic analysis rows); *Design recommendations* (torsional irregularity row).

Left-margin label: TU-TS structures with stiffness eccentricity and re-entrant corners having RI profile of irregularity

Consideration of structural irregularity based on location-specific influence	SDC E and F	SDC E and F	N/A	N/A	SDC E and F	–
All possible directions of seismic excitation (critical response)	All SDC	All SDC	N/A	N/A	All SDC	–
Detailing	Collectors are recommended to be used for SDC B, C, D, E and F to cater with excessive tensile deformations at the FS of the structure and to facilitate force transfer mechanism					
Recommendations for the design of new irregular structures	Re-entrant corners as a measure of irregularity based on $C_i \leq 15\%$ is not a sensitive trigger for the expected damage response of the structure					
Recommendations for the health monitoring of existing irregular structures	The contour plots presented in this research are proposed to be incorporated for expected weak locations in existing irregular structures					

Table 8.7 TS-IRI structures with re-entrant corners

TS structures with stiffness eccentricities and re-entrant corners having IRI profile of irregularities

Type of irregularity	$(KI_1^{s-1} + MI_2^{s-1})_P$	$(KI_1^{(s-1)})_P + (MI_2^{(s-1)})_V$	$SI_3^{s-1}\ V$	$WI_4^{s-1}\ V$	CI_5^{s-1}	$BI_6^{s-1}\ P$
Quantitative/qualitative trigger	Displaced C_R and C_M from C_G	Displaced C_R from C_G and $M_i < 1.5\ M_a$	$K_i < 0.7K_{i+1}$	$K_i < 0.65K_{i+1}$	$C_i \leq 15\%$	Converged C_M and C_R but dislocated from C_G
Seismic risk triggers for various irregularities						
Global seismic risk level						
Location-specific seismic risk level/FS			N/A	N/A		
Location-specific seismic risk level/SS			N/A	N/A		
Floor-specific seismic risk level (floor containing eccentricity)			N/A	N/A		
Irregularity influence at the adjacent floors when the irregularity is in the lower floor			N/A	N/A	N/A	N/A
Irregularity influence at the adjacent floors when the irregularity is at the top roof level floor			N/A	N/A	N/A	N/A
Remarks	Re-entrant corners as a measure of irregularity based on $C_i \leq 15\%$ is a justifiable quantitative trigger for the expected damage response in such kind of structure					
Recommended analysis procedure						
3D analysis recommendation	All SDC	All SDC	N/A	N/A	All SDC	All SDC
Prohibition of equivalent lateral force method	SDC E and F	SDC E and F	N/A	N/A	SDC E and F	–
Orthogonality influence	All SDC	All SDC	N/A	N/A	All SDC	–
Linear/nonlinear dynamic analysis	SDC E and F	SDC E and F	N/A	N/A	SDC E and F	–
Design recommendations						
Torsional irregularity based on rotation instead of drift demand	SDC E and F	SDC E and F	N/A	N/A	SDC E and F	–

Consideration of structural irregularity based on location-specific influence	All SDC	All SDC	N/A	N/A	All SDC	–
All possible directions of seismic excitation (critical response)	All SDC	All SDC	N/A	N/A	All SDC	All SDC
Detailing	Collectors are recommended to be used for SDC type B, C, D, E and F structures to cater with excessive tensile deformations at the FS of the structure and to facilitate force transfer mechanism					
Recommendations for the design of new irregular structures	Strictly avoiding the presented irregularities for SDC E and F structures					
Recommendations for the health monitoring of existing irregular structures	The contour plots presented in this research are proposed to be incorporated for expected weak locations in existing irregular structures					

Table 8.8 TF-IRI structures with soft/weak floor

Type of irregularity		$(Kl_1^{s-1} + Ml_2^{s-1})_P$	$(Kl_1^{s-1})_P + (Vl_2^{s-1})_P$	$Sl_3^{s-1}\ V$	$Wl_4^{s-1}\ V$	Cl_5^{s-1}	$Bl_6^{s-1}\ P$
	Quantitative/qualitative trigger	Displaced C_R and C_M from C_G	Displaced C_R and C_V from C_G	$K_i < 0.7K_{i+1}$	$K_i < 0.65K_{i+1}$	$C_i \leq 15\%$	Converged C_M and C_R but dislocated from C_G
Seismic risk triggers for various irregularities	Global seismic risk level					N/A	
	Location-specific seismic risk level/FS					N/A	
	Location-specific seismic risk level/SS					N/A	
	Floor-specific seismic risk level (floor containing eccentricity)					N/A	
	Irregularity influence at the adjacent floors when the irregularity is in the lower floor				N/A	N/A	
	Irregularity influence at the adjacent floors when the irregularity is at the top roof level floor				N/A	N/A	
	Remarks	Code defined mass irregularity is found to be highly non-conservative for both local and global seismic response of the floor containing irregularity especially if the vertical mass eccentricity is present in a weak or soft story. If the vertical mass irregularity is present on a specific floor and the corresponding floor is also weak and/or soft floor, the proposed quantitative trigger for vertical mass irregularity based on this research is recommended as $M_i < 1.2\ M_a$.					
Recommended analysis procedure	3D analysis recommendation	All SDC	All SDC	All SDC	All SDC	N/A	All SDC
	Prohibition of equivalent lateral force method	SDC B through F	SDC B through F	SDC B through F	SDC B through F	N/A	SDC B through F
	Orthogonality influence	All SDC	All SDC	All SDC	All SDC	N/A	All SDC
	Linear/nonlinear dynamic analysis	SDC B through F	SDC B through F	SDC B through F	SDC B through F	N/A	SDC B through F

TF structures containing mass, stiffness and strength eccentricities on the soft/weak floor of the irregular structure having IRI profile of irregularity

Design recommendations						
	All SDC	All SDC	All SDC	All SDC	N/A	All SDC
Torsional irregularity based on rotation instead of drift demand	All SDC	All SDC	All SDC	All SDC	N/A	All SDC
Consideration of structural irregularity based on location-specific influence	SDC B through F	SDC B through F	SDC B through F	SDC B through F	SDC B through F	SDC B through F
All possible directions of seismic excitation (critical response)	SDC B through F	SDC B through F	SDC B through F	SDC B through F	N/A	SDC B through F
Detailing	Collectors are recommended to be used for SDC B through F to cater with excessive tensile deformations at the FS of the structure and to facilitate force transfer mechanism. Moreover, the FS frame of the structure is recommended to be braced. The SS of the structure also requires seismic resisting detailing during the design process as the structure experienced higher tensile deformations at the SS on the floor corresponding to irregularity compared with the FS of the structure.					
Recommendations for the design of new irregular structures	Strictly avoiding the presented irregularities for SDC type D, E and F structures					
Recommendations for the health monitoring of existing irregular structures	The contour plots presented in this research are proposed to be incorporated for expected weak locations in existing irregular structures					

are highly sensitive towards seismic damage at both the FS and SS of the structure. The following suggestions are recommended for the design process of such kind of structures:

- Code-based vertical mass irregularity should be considered $M_i < 1.2\ M_a$ when the weak/soft floor is present in the structure as the interaction of irregularities leads to highly enhanced response under seismic excitations.
- ELF method of analysis should be prohibited for the SDC B, C, D, E and F as the method is based only on the fundamental mode of the structure and cannot consider complex interaction of irregularities under seismic excitations.
- It is suggested to use critical seismic response based on seismic excitation along all possible directions of the lateral moment resisting systems for SDC B through F.

8.5   Summary

Based on the findings of this research, it can be concluded that code-based quantitative triggers for structural irregularities are slightly over-conservative for circumstances when each type of irregularity is present solely in an asymmetric structure. However, when an irregular structure contains various irregularities simultaneously, the seismic response becomes highly enhanced. For instance, mass irregularity exceeding the quantitative trigger of $M_i < 1.5\ M_a$ doesn't have a pronounced influence on the seismic response. However, when the same irregularity is present on the soft/weak story, both local and global damage responses are significantly amplified. This happens due to the interaction of different irregularities present simultaneously in an asymmetric structure. Considering the interaction of irregularities, the code defined irregularity triggers are non-conservative and in such cases the safety of the asymmetric structures becomes questionable. It should also be noted that modern structures in the current times contain various irregularities if not all, simultaneously and therefore, the code defined triggers are incapable of measuring the actual irregularity in the structure for the presented cases in this chapter. To deal with this issue, it is suggested in this book to use seismic rotational demands at the FS and SS, which measure more genuinely the degree of irregularity compared with the seismic drift demands. Under increased torsional vibrations, both the flexible and stiff edge of the asymmetric structure may experience differential seismic rotational demands. Such seismic rotational demands at the individual edge have strong relevance with the floor twisting of the structure as has been explained in Chapter 6. Therefore, edge rotations are considered a significant measure to estimate the probability of the local failure as well as the global torsional behavior threatening the robustness of the asymmetric structure. This is mainly for a fact that asymmetric systems, which experience differential twisting at the edge-elements, undergo in-plane bending because the relative stiffness of the horizontal to vertical structural systems affects the torsional resistance of the frames and the in-plane rotation of the slabs. Furthermore, it is suggested in this chapter to avoid equivalent lateral force (ELF) method for the analysis of highly torsional structures. The reason for suggesting this prohibition is that ELF method of analysis is based on the fundamental mode of the structure. However, in these kinds of structures, the translational mode is coupled with the torsional mode eventually leading to different seismic demands. Moreover, under extreme torsional cases, the orthogonality influence is very high; therefore, it is proposed that under such circumstances, the design of asymmetric structures should be based on critical seismic demands, which is produced when seismic excitations are considered along with all possible directions.

Chapter 9

Conclusions

9.1 Local damage behavior of asymmetric structures

The following conclusions are established from the local damage response:

- Asymmetric reinforced concrete structures with in-plan stiffness eccentricity are observed to experience stiffness reduction on the flexible edge even when the structure is in the elastic state. This stiffness reduction is attributed to the micro-cracking state, which has a significant impact on the seismic design of the asymmetric structures as the seismic response may significantly alter because of varying dynamic characteristics of the structure and structural correspondence with the frequency band of the input acceleration. In case of visually no physical damage to the structure, the possibility for any potential damage within the concrete can be assessed in advance from residual strain data along with the characteristics of dynamic properties of the structure.
- The design of the corner members on the FS of the asymmetric structures need higher design redundancy and special seismic detailing as the torsional response may influence such members the most in asymmetric structures with in-plan stiffness eccentricity.
- In bi-eccentric TS-RI, bi-eccentric TS-IRI, mono-eccentric TS-RI and mono-eccentric TS-IRI structures, the first floor is likely to experience equally negligible compressive and tensile strains whereas the top roof is likely to experience higher tensile deformations at the FS of the structure as in such of kind of asymmetric structures, the FS of the structure is more sensitive towards tensile deformations.
- In bi-eccentric TS-IRI structures with strength eccentricities, it can be said that the structure is likely to experience negligible compressive strains at the first-floor and top roof levels. However, higher tensile and compressive strains are expected at the SS of the intermediate floor under the influence of second mode dominance.
- In bi-eccentric TF-IRI structures with strength eccentricities, highly abnormal and unpredictable damage behavior may occur at all floor levels because of the higher torsional flexibility of the structure. For such structures, it was observed that the first floor experienced higher absolute tensile deformations at the FS compared with SS of the structure whereas the SS of the structure experienced higher compressive deformations compared with the FS.

9.2   Global behavior of asymmetric structures

The conclusions established from the global behavior of asymmetric structures are as follows:

- From a global response perspective, to identify the ultimate state of collapse of a structure and structural performance, the horizontal displacement is primarily used as a threshold. The current experimental findings for the damaged state of RC model suggest that it is reasonable to expect a stable frame behavior if the following two conditions are met: (a) top roof drift is below 2.7%, and (b) inter-story drift is below 1.5% at any floor. Therefore, based on the structural damage response, top-story displacement and inter-story drift response, the top-drift has been proposed to be limited to 2% and the inter-story drift has been proposed to be limited to 1.5% as a general measure of stability for asymmetric structures with in-plan stiffness eccentricity. Moreover, displacement responses are found to be less sensitive to the local oscillation behavior as compared to the acceleration responses. This is due to the fact that low-frequency global vibrations induced through the input excitation are expected to dominate the main wave-form and can cause the local oscillations to produce additional accelerations of high frequency with local effects.
- In TS-IRI structures, the top floor is expected to experience the highest influence of translational demands when the eccentricities are at the lower floor levels. Conversely, the lower floor eccentricity has the highest influence on the top floor translational demands.
- Regardless of the type of asymmetry, seismic translational demands are likely to decrease at both the FS and SS of the structure when C_M and C_R are converged at one point but dislocated from the C_G of the structure.
- Strength eccentricity appears to be more sensitive than the stiffness eccentricity and is expected to induce high translational and rotational demands. Moreover, structures with strength eccentricities and IRI state of irregularity produce very unpredictable and abnormal response both in terms of translational and rotational demands.
- From a statistical viewpoint, comparing the relative displacement, velocity and acceleration responses, structural response sensitivity is more towards acceleration response than the velocity and displacement response. The sensitivity of the acceleration response was observed to be higher at the FS than the SS. Comparing the same response quantities in terms of response variability, acceleration response has the higher response variability than other response quantities. However, the acceleration variability greatly influenced the FS with respect to the SS of the structure. Moreover, treating the experimental orientation as a randomly selected direction, there exists approximately 80% probability that for most of the rotational response quantities, the seismic response will exceed the experimental orientation's response. Besides, for most of the relative response quantities, the probability of exceedance appeared to be lower at SS compared to the response at the FS of the structure.

References

ABAQUS (2003) *ABAQUS/CAE User's Manual: Version 6.4*. ABAQUS, Pawtucket, RI.

ACI-318 (2008) *Building code requirements for structural concrete*: (ACI 318-08); and Commentary (ACI 318R-08). Farmington Hills, MI: American Concrete Institute.

Adalier, K. & Aydingun, O. (1998) Adana-Ceyhan (Turkey) Earthquake: Engineering Reconnaissance Report. Technical Report No. Geo98–9, Department of Civil Engineering, Eastern Mediterranean University, TRN Cyprus, June 27, 45p. Google Scholar.

Alam, Z., Zhang, C., & Samali, B. (2017) Response uncertainty under varying orientations of ground motions. In *Mechanics of Structures and Materials: Advancements and Challenges*. Proceedings of the 24th Australasian Conference on the Mechanics of Structures and Materials (ACMSM24), Perth, Australia, 6–9 December 2016 (pp. 657–662).

Anagnostopoulos, S.A., Alexopoulou, C. & Stathopoulos, K.G. (2010) An answer to an important controversy and the need for caution when using simple models to predict inelastic earthquake response of buildings with torsion. *Earthquake Engineering & Structural Dynamics*, 39, 521–540.

Anagnostopoulos, S.A., Kyrkos, M., Papalymperi, A. & Plevri, E. (2015a) Should accidental eccentricity be eliminated from Eurocode 8. *Earthquakes and Structures*, 8, 463–484.

Anagnostopoulos, S.A., Kyrkos, M. & Stathopoulos, K. (2015b) Earthquake induced torsion in buildings: Critical review and state of the art. *Earthquakes and Structures*, 8, 305–377.

Arsenault, T.J., Achuthan, A., Marzocca, P., Grappasonni, C. & Coppotelli, G. (2013) Development of a FBG based distributed strain sensor system for wind turbine structural health monitoring. *Smart Materials and Structures*, 22, 075027.

ASCE/SEI (2016) *Minimum Design Loads for Buildings and Other Structures*, American Society of Civil Engineers (ASCE) Reston, VA. https://doi.org/10.1061/9780784414248.

Athanassiadou, C. (2008) Seismic performance of R/C plane frames irregular in elevation. *Engineering Structures*, 30, 1250–1261.

Athanatopoulou, A. (2005) Critical orientation of three correlated seismic components. *Engineering Structures*, 27, 301–312.

Aziminejad, A. & Moghadam, A. (2009) Performance of asymmetric multistory shear buildings with different strength distributions. *Journal of Applied Sciences*, 9, 1082–1089.

Bang, H.-J., Kim, H.-I. & Lee, K.-S. (2012) Measurement of strain and bending deflection of a wind turbine tower using arrayed FBG sensors. *International Journal of Precision Engineering and Manufacturing*, 13, 2121–2126.

Barbosa, C., Costa, N., Ferreira, L., Araújo, F., Varum, H., Costa, A., Fernandes, C. & Rodrigues, H. (2008) Weldable fibre Bragg grating sensors for steel bridge monitoring. *Measurement Science and Technology*, 19, 125305.

Basu, D. & Giri, S. (2015) Accidental eccentricity in multistory buildings due to torsional ground motion. *Bulletin of Earthquake Engineering*, 13, 3779–3808.

Bathe, K.-J. (2006) *Finite Element Procedures*. Englewood Cliffs, NJ: Prentice-Hall.

Benavent-Climent, A., Morillas, L. & Escolano-Margarit, D. (2014) Inelastic torsional seismic response of nominally symmetric reinforced concrete frame structures: Shaking table tests. *Engineering Structures*, 80, 109–117.

Bhatt, C. & Bento, R. (2014) The extended adaptive capacity spectrum method for the seismic assessment of plan-asymmetric buildings. *Earthquake Spectra*, 30, 683–703.

Bikçe, M. & Çelik, T.B. (2016) Failure analysis of newly constructed RC buildings designed according to 2007 Turkish Seismic Code during the October 23, 2011 Van earthquake. *Engineering Failure Analysis*, 64, 67–84.

Biswas, P., Bandyopadhyay, S., Kesavan, K., Parivallal, S., Sundaram, B.A., Ravisankar, K. & Dasgupta, K. (2010) Investigation on packages of fiber Bragg grating for use as embeddable strain sensor in concrete structure. *Sensors and Actuators A: Physical*, 157, 77–83.

Bosco, M., Marino, E. & Rossi, P.P. (2004) Limits of application of simplified design procedures to non-regularly asymmetric buildings. *13th World Conference on Earthquake Engineering*. Vancouver, British Columbia, Canada, August 1–6.

Bousias, S., Fardis, M., Spathis, A.L. & Kosmopoulos, A. (2007) Pseudodynamic response of torsionally unbalanced two-story test structure. *Earthquake Engineering & Structural Dynamics*, 36, 1065–1087.

Bozorgnia, Y. & Tso, W. (1986) Inelastic earthquake response of asymmetric structures. *Journal of Structural Engineering*, 112, 383–400.

Brönnimann, R., Nellen, P.M. & Sennhauser, U. (1998) Application and reliability of a fiber optical surveillance system for a stay cable bridge. *Smart Materials and Structures*, 7, 229.

Cao, W., Zhao, C., Xue, S. & Zhang, J. (2009) Shaking table experimental study of the short pier RC shear wall structures with concealed bracing. *Advances in Structural Engineering*, 12, 267–278.

Chan, P.K., Jin, W., Lau, A.K. & Zhou, L. (2000) Strain monitoring of composite-boned concrete specimen measurements by use of FMCW multiplexed fiber Bragg grating sensor array. *International Conference on Sensors and Control Techniques (ICSC2000), International Society for Optics and Photonics*. Wuhan, China, pp. 56–59.

Chan, T.H., Yu, L., Tam, H.-Y., Ni, Y.-Q., Liu, S., Chung, W. & Cheng, L. (2006) Fiber Bragg grating sensors for structural health monitoring of Tsing Ma bridge: Background and experimental observation. *Engineering Structures*, 28, 648–659.

Chandler, A. & Duan, X. (1991) Evaluation of factors influencing the inelastic seismic performance of torsionally asymmetric buildings. *Earthquake Engineering & Structural Dynamics*, 20, 87–95.

Chandler, A. & Hutchinson, G. (1986) Torsional coupling effects in the earthquake response of asymmetric buildings. *Engineering Structures*, 8, 222–236.

Chandler, A. & Hutchinson, G. (1992) Effect of structural period and ground motion parameters on the earthquake response of asymmetric buildings. *Engineering Structures*, 14, 354–360.

Chandler, A., Correnza, J. & Hutchinson, G. (1995) Influence of accidental eccentricity on inelastic seismic torsional effects in buildings. *Engineering Structures*, 17, 167–178.

Chen, Q.-F. & Wang, K. (2010) The 2008 Wenchuan earthquake and earthquake prediction in China. *Bulletin of the Seismological Society of America*, 100, 2840–2857.

Chintanapakdee, C. & Chopra, A.K. (2004) Seismic response of vertically irregular frames: Response history and modal pushover analyses. *Journal of Structural Engineering*, 130, 1177–1185.

Chopra, A.K. (2001) *Dynamics of Structures: Theory and Applications to Earthquake Engineering*. Englewood Cliffs, NJ: Prentice-Hall.

Chopra, A.K. & Chintanapakdee, C. (2004) Evaluation of modal and FEMA pushover analyses: Vertically "regular" and irregular generic frames. *Earthquake Spectra*, 20, 255–271.

Chopra, A.K. & Goel, R.K. (1991) Evaluation of torsional provisions in seismic codes. *Journal of Structural Engineering*, 117, 3762–3782.

Chopra, A.K. & Goel, R.K. (2004) A modal pushover analysis procedure to estimate seismic demands for unsymmetric-plan buildings. *Earthquake Engineering & Structural Dynamics*, 33, 903–927.

Chung, W., Kim, S., Kim, N.-S. & Lee, H.-U. (2008) Deflection estimation of a full scale prestressed concrete girder using long-gauge fiber optic sensors. *Construction and Building Materials*, 22, 394–401.

Cimellaro, G.P., Giovine, T. & Lopez-Garcia, D. (2014) Bidirectional pushover analysis of irregular structures. *Journal of Structural Engineering*, 140, 04014059.

Costa, B.J.A. & Figueiras, J.A. (2012) Fiber optic based monitoring system applied to a centenary metallic arch bridge: Design and installation. *Engineering Structures*, 44, 271–280.

D'ambrisi, A., De Stefano, M. & Tanganelli, M. (2009) Use of pushover analysis for predicting seismic response of irregular buildings: A case study. *Journal of Earthquake Engineering*, 13, 1089–1100.

Das, S. & Nau, J.M. (2003) Seismic design aspects of vertically irregular reinforced concrete buildings. *Earthquake Spectra*, 19, 455–477.

De-La-Colina, J. (1999a) Effects of torsion factors on simple non-linear systems using fully-bidirectional analyses. *Earthquake Engineering & Structural Dynamics*, 28, 691–706.

De-La-Colina, J. (1999b) In-plane floor flexibility effects on torsionally unbalanced systems. *Earthquake Engineering & Structural Dynamics*, 28, 1705–1715.

De-La-Colina, J. (2003) Assessment of design recommendations for torsionally unbalanced multistory buildings. *Earthquake Spectra*, 19, 47–66.

De-La-Colina, J., Acuña, Q., Hernández, A. & Valdés, J. (2007) Laboratory tests of steel simple torsionally unbalanced models. *Earthquake Engineering & Structural Dynamics*, 36, 887–907.

Dempsey, K. & Tso, W. (1982) An alternative path to seismic torsional provisions. *International Journal of Soil Dynamics and Earthquake Engineering*, 1, 3–10.

De Stefano, M. & Pintucchi, B. (2002) A model for analyzing inelastic seismic response of plan-irregular building structures. *Proceedings of the 15th ASCE Engineering Mechanics Conference, CD ROM, New York*.

De Stefano, M., Faella, G. & Ramasco, R. (1998) Inelastic seismic response of one-way plan-asymmetric systems under bi-directional ground motions. *Earthquake Engineering & Structural Dynamics*, 27, 363–376.

De Stefano, M., Marino, E.M. & Rossi, P.P. (2006) Effect of overstrength on the seismic behaviour of multi-story regularly asymmetric buildings. *Bulletin of Earthquake Engineering*, 4, 23–42.

Dimova, S.L. & Alashki, I. (2003) Seismic design of symmetric structures for accidental torsion. *Bulletin of Earthquake Engineering*, 1, 303–320.

Du, K., Ding, B., Luo, H. & Sun, J. (2017) Relationship between Peak Ground Acceleration, Peak Ground Velocity, and Macroseismic Intensity in Western China relationship between PGA, PGV, and Macroseismic Intensity in Western China. *Bulletin of the Seismological Society of America*, 109(1), 284–297.

Duan, X. & Chandler, A. (1997) An optimized procedure for seismic design of torsionally unbalanced structures. *Earthquake Engineering & Structural Dynamics*, 26, 737–757.

Dutta, S. & Das, P. (2002a) Inelastic seismic response of code-designed reinforced concrete asymmetric buildings with strength degradation. *Engineering Structures*, 24, 1295–1314.

Dutta, S. & Das, P. (2002b) Validity and applicability of two simple hysteresis models to assess progressive seismic damage in R/C asymmetric buildings. *Journal of Sound and Vibration*, 257, 753–777.

EC-8 (2005) *Design of Structures for Earthquake Resistance-Part 1: General Rules, Seismic Actions and Rules for Buildings*. European Committee for Standardization, Brussels.

Eivani, H., Moghadam, A.S., Aziminejad, A. & Nekooei, M. (2018) Seismic response of plan-asymmetric structures with diaphragm flexibility. *Shock and Vibration*. https://doi.org/10.1155/2018/4149212.

Elwood, K.J. & Moehle, J.P. (2008a) Dynamic collapse analysis for a reinforced concrete frame sustaining shear and axial failures. *Earthquake Engineering & Structural Dynamics*, 37, 991–1012.

Elwood, K.J. & Moehle, J.P. (2008b) Dynamic shear and axial-load failure of reinforced concrete columns. *Journal of Structural Engineering*, 134, 1189–1198.

Fajfar, P., Marušić, D. & Peruš, I. (2005) Torsional effects in the pushover-based seismic analysis of buildings. *Journal of Earthquake Engineering*, 9, 831–854.

Fang, C.-H. & Leon, R.T. (2018) Seismic behavior of symmetric and asymmetric steel structures with rigid and semirigid diaphragms. *Journal of Structural Engineering*, 144, 04018186.

Ferhi, A. & Truman, K. (1996a) Behaviour of asymmetric building systems under a monotonic load: I. *Engineering Structures*, 18, 133–141.

Ferhi, A. & Truman, K. (1996b) Behaviour of asymmetric building systems under a monotonic load: II. *Engineering Structures*, 18, 142–153.

Fujii, K., Nakano, Y. & Sanada, Y. (2004) A simplified nonlinear analysis procedure for single-story asymmetric buildings. *Journal of Japan Association for Earthquake Engineering*, 4, 1–20.

GB17742 (2008) *The Chinese seismic intensity scale*. Administration of Quality Supervision, Inspection and Quarantine of People's Republic of China (AQSIQ), Beijing, China (English Translation).

GB50011-2001 (2001) *Code for Seismic Design of Buildings (GB50011-2001)*. China Architecture and Building Press, Beijing (English Translation).

Georgoussis, G.K. (2016) An approach for minimum rotational response of medium-rise asymmetric structures under seismic excitations. *Advances in Structural Engineering*, 19, 420–436.

Ger, J.-F., Cheng, F.Y. & Lu, L.-W. (1993) Collapse behavior of Pino Suarez building during 1985 Mexico City earthquake. *Journal of Structural Engineering*, 119, 852–870.

Ghannoum, W.M. & Moehle, J.P. (2012) Shake-table tests of a concrete frame sustaining column axial failures. *ACI Structural Journal*, 109, 393.

Ghersi, A. & Rossi, P.P. (2001) Influence of bi-directional ground motions on the inelastic response of one-story in-plan irregular systems. *Engineering Structures*, 23, 579–591.

Gilbert, R.I. & Warner, R.F. (1978) Tension stiffening in reinforced concrete slabs. *Journal of the Structural Division*, 104, 1885–1900.

Goel, R.K. & Chopra, A.K. (1990) Inelastic seismic response of one-story, asymmetric-plan systems: Effects of stiffness and strength distribution. *Earthquake Engineering & Structural Dynamics*, 19, 949–970.

Harasimowicz, A.P. & Goel, R.K. (1998) Seismic code analysis of multi-story asymmetric buildings. *Earthquake Engineering & Structural Dynamics*, 27, 173–185.

Heerema, P., Shedid, M. & El-Dakhakhni, W. (2014) Seismic response analysis of a reinforced concrete block shear wall asymmetric building. *Journal of Structural Engineering*, 141, 04014178.

Hejal, R. & Chopra, A.K. (1989a) Earthquake response of torsionally coupled, frame buildings. *Journal of Structural Engineering*, 115, 834–851.

Hejal, R. & Chopra, A.K. (1989b) Lateral-torsional coupling in earthquake response of frame buildings. *Journal of Structural Engineering*, 115, 852–867.

Heredia-Zavoni, E. & Machicao-Barrionuevo, R. (2004) Response to orthogonal components of ground motion and assessment of percentage combination rules. *Earthquake Engineering & Structural Dynamics*, 33, 271–284.

Hsu, L. & Hsu, C.-T. (1994) Complete stress-strain behaviour of high-strength concrete under compression. *Magazine of Concrete Research*, 46, 301–312.

Huang, S. & Skokan, M. (2002) Collapse of the Tungshing building during the 1999 Chi-Chi earthquake in Taiwan. *Proc., 7th US National Conf. on Earthquake Engineering*, Boston, USA.

Humar, J. & Kumar, P. (1998a) Torsional motion of buildings during earthquakes, I: Elastic response. *Canadian Journal of Civil Engineering*, 25, 898–916.

Humar, J. & Kumar, P. (1998b) Torsional motion of buildings during earthquakes, II: Inelastic response. *Canadian Journal of Civil Engineering*, 25, 917–934.

IBC (2009) International building code. *International Code Council, Inc. (Formerly BOCA, ICBO and SBCCI)*, 4051, 60478–5795.

Inel, M. & Ozmen, H.B. (2006) Effects of plastic hinge properties in nonlinear analysis of reinforced concrete buildings. *Engineering Structures*, 28, 1494–1502.

Jarernprasert, S. & Bazan, E. (2008) Inelastic torsional single-story systems. *14th World Conference on Earthquake Engineering*, Beijing, China.

Jarernprasert, S., Bazan, E. & Bielak, J. (2008) On the seismic design of inelastic asymmetric systems. *The 14th World Conference on Earthquake Engineering*. pp. 14–17, Beijing, China.

Jeong, S.H. (2005) *Experimental and Analytical Seismic Assessment of Buildings with Plan Irregularities*. Champaign, IL: University of Illinois at Urbana-Champaign.

Kalkan, E. & Kwong, N.S. (2013) Pros and cons of rotating ground motion records to fault-normal/parallel directions for response history analysis of buildings. *Journal of Structural Engineering*, 140, 04013062.

Kan, C.L. & Chopra, A.K. (1981a) Simple model for earthquake response studies of torsionally coupled buildings. *Journal of the Engineering Mechanics Division*, 107, 935–951.

Kan, C.L. & Chopra, A.K. (1981b) Torsional coupling and earthquake response of simple elastic and inelastic systems. *Journal of the Structural Division*, 107, 1569–1588.

Karavasilis, T.L., Bazeos, N. & Beskos, D. (2008) Seismic response of plane steel MRF with setbacks: Estimation of inelastic deformation demands. *Journal of Constructional Steel Research*, 64, 644–654.

Karayannis, C.G. & Naoum, M.C. (2018) Torsional behavior of multistory RC frame structures due to asymmetric seismic interaction. *Engineering Structures*, 163, 93–111.

Kerrouche, A., Boyle, W., Gebremichael, Y., Sun, T., Grattan, K., Täljsten, B. & Bennitz, A. (2008a) Field tests of fibre Bragg grating sensors incorporated into CFRP for railway bridge strengthening condition monitoring. *Sensors and Actuators A: Physical*, 148, 68–74.

Kerrouche, A., Leighton, J., Boyle, W., Gebremichael, Y., Sun, T., Grattan, K.T. & Taljsten, B. (2008b) Strain measurement on a rail bridge loaded to failure using a fiber Bragg grating-based distributed sensor system. *IEEE Sensors Journal*, 8, 2059–2065.

Kerrouche, A., Boyle, W., Sun, T. & Grattan, K. (2009) Design and in-the-field performance evaluation of compact FBG sensor system for structural health monitoring applications. *Sensors and Actuators A: Physical*, 151, 107–112.

Kersey, A.D., Davis, M.A., Patrick, H.J., Leblanc, M., Koo, K., Askins, C., Putnam, M. & Friebele, E.J. (1997) Fiber grating sensors. *Journal of Lightwave Technology*, 15, 1442–1463.

Kim, S.-W., Kang, W.-R., Jeong, M.-S., Lee, I. & Kwon, I.-B. (2013) Deflection estimation of a wind turbine blade using FBG sensors embedded in the blade bonding line. *Smart Materials and Structures*, 22, 125004.

Kim, Y., Kabeyasawa, T. & Igarashi, S. (2012) Dynamic collapse test on eccentric reinforced concrete structures with and without seismic retrofit. *Engineering Structures*, 34, 95–110.

Kister, G., Winter, D., Gebremichael, Y., Leighton, J., Badcock, R., Tester, P., Krishnamurthy, S., Boyle, W., Grattan, K. & Fernando, G. (2007) Methodology and integrity monitoring of foundation concrete piles using Bragg grating optical fibre sensors. *Engineering Structures*, 29, 2048–2055.

Kostinakis, K.G., Athanatopoulou, A.M. & Avramidis, I.E. (2008) Maximum response and critical incident angle in special classes of buildings subjected to two horizontal seismic components. In *Proceedings of the 6th GRACM International Congress on Computational Mechanics* (No. IKEECONF-2018-333). Aristotle University of Thessaloniki.

Kyrkos, M. & Anagnostopoulos, S. (2011) An assessment of code designed, torsionally stiff, asymmetric steel buildings under strong earthquake excitations. *Earthquake and Structures*, 2, 109–126.

Ladinovic, D. (2008) Nonlinear seismic analysis of asymmetric in plan building. *Facta Universitatis: Architecture and Civil Engineering*, 6.

Ladjinovic, D., Raseta, A., Radujkovic, A., Folic, R. & Prokic, A. (2012) Comparison of structural models for seismic analysis of multi-story frame buildings. *World Conference on Earthquake Engineering (15WCEE)*, Lisbon, Portugal.

Lee, H.-S. & Hwang, K.R. (2015) Torsion design implications from shake-table responses of an RC low-rise building model having irregularities at the ground story. *Earthquake Engineering & Structural Dynamics*, 44, 907–927.

Lee, H.-S. & Ko, D.-W. (2004) Seismic response of high-rise RC bearing-wall structures with irregularities at bottom stories. *Proceedings of the 13th World Conference on Earthquake Engineering, Vancouver, Canada.*

Lee, H.-S. & Woo, S.-W. (2002) Seismic performance of a 3-story RC frame in a low-seismicity region. *Engineering Structures*, 24, 719–734.

Lee, H.-S., Jung, D.W., Lee, K.B., Kim, H.C. & Lee, K. (2011) Shake-table responses of a low-rise RC building model having irregularities at first story. *Structural Engineering and Mechanics*, 40, 517–539.

Lee, J. & Fenves, G.L. (1998) Plastic-damage model for cyclic loading of concrete structures. *Journal of Engineering Mechanics*, 124, 892–900.

Li, C., Lam, S., Zhang, M. & Wong, Y. (2006) Shaking table test of a 1: 20 scale high-rise building with a transfer plate system. *Journal of Structural Engineering*, 132, 1732–1744.

Li, C., Zhao, Y.-G., Liu, H., Wan, Z., Zhang, C. & Rong, N. (2008) Monitoring second lining of tunnel with mounted fiber Bragg grating strain sensors. *Automation in Construction*, 17, 641–644.

Li, C., Zhao, Y.-G., Liu, H., Wan, Z., Xu, J.-C., Xu, X.-P. & Chen, Y. (2009) Strain and back cavity of tunnel engineering surveyed by FBG strain sensors and geological radar. *Journal of Intelligent Material Systems and Structures*, 20, 2285–2289.

Li, D.S., Ren, L., Li, H.-N. & Song, G. (2012) Structural health monitoring of a tall building during construction with fiber Bragg grating sensors. *International Journal of Distributed Sensor Networks*, 8, 272190.

Li, H.-N., Li, D.-S. & Song, G.-B. (2004) Recent applications of fiber optic sensors to health monitoring in civil engineering. *Engineering Structures*, 26, 1647–1657.

Li, H.-N., Sun, L. & Song, G. (2004) Modal combination method for earthquake-resistant design of tall structures to multidimensional excitations. *The Structural Design of Tall and Special Buildings*, 13, 245–263.

Li, S., Zuo, Z., Zhai, C., Xu, S. & Xie, L. (2016) Shaking table test on the collapse process of a three-story reinforced concrete frame structure. *Engineering Structures*, 118, 156–166.

Lignos, D.G. & Gantes, C.J. (2005) Seismic demands for steel braced frames with stiffness irregularities based on modal pushover analysis. *Proceedings of the 4th European Workshop on the Seismic Behaviour of Irregular and Complex Structures,* CD ROM, Thessaloniki.

Lignos, D.G. & Krawinkler, H. (2010) Deterioration modeling of steel components in support of collapse prediction of steel moment frames under earthquake loading. *Journal of Structural Engineering*, 137, 1291–1302.

Lim, H.-K., Kang, J.W., Pak, H., Chi, H.-S., Lee, Y.-G. & Kim, J. (2018) Seismic response of a three-dimensional asymmetric multi-story reinforced concrete structure. *Applied Sciences*, 8, 479.

Lin, Y.-B., Chen, J.-C., Chang, K.-C., Chern, J.-C. & Lai, J.-S. (2005a) Real-time monitoring of local scour by using fiber Bragg grating sensors. *Smart Materials and Structures*, 14, 664.

Lin, Y.-B., Pan, C.L., Kuo, Y.H., Chang, K.C. & Chern, J.C. (2005b) Online monitoring of highway bridge construction using fiber Bragg grating sensors. *Smart Materials and Structures*, 14, 1075.

Lin, Y.-B., Lai, J.S., Chang, K.C. & Li, L.S. (2006) Flood scour monitoring system using fiber Bragg grating sensors. *Smart Materials and Structures*, 15, 1950.

Liu, J., Liu, F., Kong, X. & Yu, L. (2016) Large-scale shaking table model tests on seismically induced failure of concrete-faced rockfill dams. *Soil Dynamics and Earthquake Engineering*, 82, 11–23.

López, O.A. & Torres, R. (1997) The critical angle of seismic incidence and the maximum structural response. *Earthquake Engineering & Structural Dynamics*, 26, 881–894.

Lopez, O.A., Chopra, A.K. & Hernandez, J.J. (2000) Critical response of structures to multicomponent earthquake excitation. *Earthquake Engineering & Structural Dynamics*, 29, 1759–1778.

López, O.A., Chopra, A.K. & Hernández, J.J. (2001) Evaluation of combination rules for maximum response calculation in multicomponent seismic analysis. *Earthquake Engineering & Structural Dynamics*, 30, 1379–1398.

Lu, Y. (2002) Comparative study of seismic behavior of multistory reinforced concrete framed structures. *Journal of Structural Engineering*, 128, 169–178.

Lu, Z., Chen, X., Lu, X. & Yang, Z. (2016) Shaking table test and numerical simulation of an RC frame-core tube structure for earthquake-induced collapse. *Earthquake Engineering & Structural Dynamics*, 45, 1537–1556.

Lucchini, A., Monti, G. & Kunnath, S. (2010) Nonlinear response of two-way asymmetric single-story building under biaxial excitation. *Journal of Structural Engineering*, 137, 34–40.

Magliulo, G., Maddaloni, G. & Petrone, C. (2014) Influence of earthquake direction on the seismic response of irregular plan RC frame buildings. *Earthquake Engineering and Engineering Vibration*, 13, 243–256.

Majumder, M., Gangopadhyay, T.K., Chakraborty, A.K., Dasgupta, K. & Bhattacharya, D.K. (2008) Fibre Bragg gratings in structural health monitoring: Present status and applications. *Sensors and Actuators A: Physical*, 147, 150–164.

Mander, J.B., Priestley, M.J. & Park, R. (1988) Theoretical stress-strain model for confined concrete. *Journal of Structural Engineering*, 114, 1804–1826.

Marušić, D. & Fajfar, P. (2005) On the inelastic seismic response of asymmetric buildings under biaxial excitation. *Earthquake Engineering & Structural Dynamics*, 34, 943–963.

Mccrum, D.P. & Broderick, B. (2013) An experimental and numerical investigation into the seismic performance of a multi-story concentrically braced plan irregular structure. *Bulletin of Earthquake Engineering*, 11, 2363–2385.

Menun, C. & Kiureghian, A.D. (1998) A replacement for the 30%, 40%, and SRSS rules for multicomponent seismic analysis. *Earthquake Spectra*, 14, 153–163.

Menun, C. & Kiureghian, A.D. (2000a) Envelopes for seismic response vectors. I: Theory. *Journal of Structural Engineering*, 126, 467–473.

Menun, C. & Kiureghian, A.D. (2000b) Envelopes for seismic response vectors. II: Application. *Journal of Structural Engineering*, 126, 474–481.

Mexico-Code (2004) *Complementary Technical Norms for Earthquake Resistant Design*. MCBC, Mexico City, Mexico.

Minzheng, Z. & Yingjie, J. (2008) Building damage in Dujiangyan during Wenchuan earthquake. *Earthquake Engineering and Engineering Vibration*, 7, 263–269.

Mita, A. & Yokoi, I. (2001) Fiber Bragg grating accelerometer for buildings and civil infrastructures. *Proc. SPIE*, 479–486, Newport Beach, CA.

Moehle, J.P. (1984) Seismic response of vertically irregular structures. *Journal of Structural Engineering*, 110, 2002–2014.

Moehle, J.P. & Alarcon, L.F. (1986) Seismic analysis methods for irregular buildings. *Journal of Structural Engineering*, 112, 35–52.

Moghadam, A. & Tso, W. (2000) Pushover analysis for asymmetric and set-back multi-story buildings. *Proceedings of the 12th World Conference on Earthquake Engineering*, 1093, Auckland, New Zealand.

Myslimaj, B. & Tso, W. (2002) A strength distribution criterion for minimizing torsional response of asymmetric wall-type systems. *Earthquake Engineering & Structural Dynamics*, 31, 99–120.

Naruse, H., Uchiyama, Y., Kurashima, T. & Unno, S. (2000) River levee change detection using distributed fiber optic strain sensor. *IEICE Transactions on Electronics*, 83, 462–467.

Nayal, R. & Rasheed, H.A. (2006) Tension stiffening model for concrete beams reinforced with steel and FRP bars. *Journal of Materials in Civil Engineering*, 18, 831–841.

NBCC (2005) National Building Code of Canada, 2005. *National Research Council of Canada (NRCC)*, Ottawa, Canada.

Negro, P., Mola, E., Molina, F.J. & Magonette, G.E. (2004) Full-scale PSD testing of a torsionally unbalanced three-story non-seismic RC frame. *Proc., 13th World Conf. on Earthquake Engineering*, Vancouver, British Columbia.

Nezhad, M.E. & Poursha, M. (2015) Seismic evaluation of vertically irregular building frames with stiffness, strength, combined-stiffness-and-strength and mass irregularities. *Earthquakes and Structures*, 9, 353–373.

Ni, S., Li, S., Chang, Z. & Xie, L. (2013) An alternative construction of normalized seismic design spectra for near-fault regions. *Earthquake Engineering and Engineering Vibration*, 12, 351–362.

Ni, Y., Xia, Y., Liao, W. & Ko, J. (2009) Technology innovation in developing the structural health monitoring system for Guangzhou New TV Tower. *Structural Control and Health Monitoring*, 16, 73–98.

Osteraas, J. & Krawinkler, H. (1989) The Mexico earthquake of September 19, 1985: Behavior of steel buildings. *Earthquake Spectra*, 5, 51–88.

Oyguc, R., Toros, C. & Abdelnaby, A.E. (2018) Seismic behavior of irregular reinforced-concrete structures under multiple earthquake excitations. *Soil Dynamics and Earthquake Engineering*, 104, 15–32.

Park, Y.-J. (1985) *Seismic Damage Analysis and Damage-Limiting Design for R/c Structures (Earthquake, Building, Reliability, Design)*. Champaign, IL: University of Illinois at Urbana-Champaign.

Pearson, C. & Delatte, N. (2005) Ronan point apartment tower collapse and its effect on building codes. *Journal of Performance of Constructed Facilities*, 19, 172–177.

Pekau, O. & Guimond, R. (1990) Accidental torsion in yielding symmetric structures. *Engineering Structures*, 12, 98–105.

Penelis, G.G. & Kappos, A. (2005) Inelastic torsional effects in 3D pushover analysis of buildings. *Proceeding of Fourth European Workshop on the Seismic Behavior of Irregular and Complex Structures, Greece, Paper*.

Penzien, J. & Watabe, M. (1974) Characteristics of 3-dimensional earthquake ground motions. *Earthquake Engineering & Structural Dynamics*, 3, 365–373.

Peruš, I. & Fajfar, P. (2005) On the inelastic torsional response of single-story structures under bi-axial excitation. *Earthquake Engineering & Structural Dynamics*, 34, 931–941.

Raghunandan, M. & Liel, A.B. (2013) Effect of ground motion duration on earthquake-induced structural collapse. *Structural Safety*, 41, 119–133.

Rao, Y.-J. (1999) Recent progress in applications of in-fibre Bragg grating sensors. *Optics and Lasers in Engineering*, 31, 297–324.

Ren, L., Li, H.-N., Zhou, J., Li, D.-S. & Sun, L. (2006a) Health monitoring system for offshore platform with fiber Bragg grating sensors. *Optical Engineering*, 45, 084401–084401–9.

Ren, L., Li, H.-N., Zhou, J., Sun, L. & Li, D.-S. (2006b) Application of tube-packaged FBG strain sensor in vibration experiment of submarine pipeline model. *China Ocean Engineering*, 20, 155–164.

Richard, B., Cherubini, S., Voldoire, F., Charbonnel, P.-E., Chaudat, T., Abouri, S. & Bonfils, N. (2016) SMART 2013: Experimental and numerical assessment of the dynamic behavior by shaking table tests of an asymmetrical reinforced concrete structure subjected to high intensity ground motions. *Engineering Structures*, 109, 99–116.

Rigato, A.B. & Medina, R.A. (2007) Influence of angle of incidence on seismic demands for inelastic single-story structures subjected to bi-directional ground motions. *Engineering Structures*, 29, 2593–2601.

Rodrigues, C., Cavadas, F., Félix, C. & Figueiras, J. (2012) FBG based strain monitoring in the rehabilitation of a centenary metallic bridge. *Engineering Structures*, 44, 281–290.

Roufaiel, M.S. & Meyer, C. (1983) *Analysis of Damaged Concrete Frame Buildings*. Technical Report no. NSF-CEE-81-21359-1, Columbia University, New York, USA.

Ruiz-García, J., Marín, M.V. & Terán-Gilmore, A. (2014) Effect of seismic sequences in reinforced concrete frame buildings located in soft-soil sites. *Soil Dynamics and Earthquake Engineering*, 63, 56–68.

Sadek, A. & Tso, W. (1989) Strength eccentricity concept for inelastic analysis of asymmetrical structures. *Engineering Structures*, 11, 189–194.

Schulz, W.L., Conte, J.P., & Udd, E. (2001) Long-gage fiber optic Bragg grating strain sensors to monitor civil structures. In *Proceedings of the SPIE's 8th Annual International Symposium on Smart Structures and Materials*, Newport Beach, CA, USA, 30 July (pp. 56–65).

Sezen, H., Whittaker, A.S., Elwood, K.J. & Mosalam, K.M. (2003) Performance of reinforced concrete buildings during the August 17, 1999 Kocaeli, Turkey earthquake, and seismic design and construction practise in Turkey. *Engineering Structures*, 25, 103–114.

Shahrooz, B.M. & Moehle, J.P. (1990a) Evaluation of seismic performance of reinforced concrete frames. *Journal of Structural Engineering*, 116, 1403–1422.

Shahrooz, B.M. & Moehle, J.P. (1990b) Seismic response and design of setback buildings. *Journal of Structural Engineering*, 116, 1423–1439.

Shakib, H. & Ghasemi, A. (2007) Considering different criteria for minimizing torsional response of asymmetric structures under near-fault and far-fault excitations. *International Journal of Civil Engineering*, 5, 247–265.

Shakya, M. & Kawan, C.K. (2016) Reconnaissance based damage survey of buildings in Kathmandu valley: An aftermath of 7.8 Mw, 25 April 2015 Gorkha (Nepal) earthquake. *Engineering Failure Analysis*, 59, 161–184.

Sharma, K., Deng, L. & Noguez, C.C. (2016) Field investigation on the performance of building structures during the April 25, 2015, Gorkha earthquake in Nepal. *Engineering Structures*, 121, 61–74.

Smeby, W. & Der Kiureghian, A. (1985) Modal combination rules for multicomponent earthquake excitation. *Earthquake Engineering & Structural Dynamics*, 13, 1–12.

Song, J., Chu, Y.-L., Liang, Z. & Lee, G.C. (2007) Estimation of peak relative velocity and peak absolute acceleration of linear SDOF systems. *Earthquake Engineering and Engineering Vibration*, 6, 1–10.

Stathopoulos, K.G. & Anagnostopoulos, S.A. (2002) Inelastic earthquake induced torsion in buildings: Results and conclusions from realistic models. *Proceedings of the 12th European Conference on Earthquake Engineering, CD ROM,* London.

Stathopoulos, K.G. & Anagnostopoulos, S.A. (2003) Inelastic earthquake response of single-story asymmetric buildings: An assessment of simplified shear-beam models. *Earthquake Engineering & Structural Dynamics*, 32, 1813–1831.

Stathopoulos, K.G. & Anagnostopoulos, S.A. (2005) Inelastic torsion of multistory buildings under earthquake excitations. *Earthquake Engineering & Structural Dynamics*, 34, 1449–1465.

Stathopoulos, K.G. & Anagnostopoulos, S.A. (2010) Accidental design eccentricity: Is it important for the inelastic response of buildings to strong earthquakes? *Soil Dynamics and Earthquake Engineering*, 30, 782–797.

Sun, L., Hao, H., Zhang, B., Ren, X. & Li, J. (2015) Strain transfer analysis of embedded fiber Bragg grating strain sensor. *Journal of Testing and Evaluation*, 44, 2312–2320.

Takewaki, I. (2005) A comprehensive review of seismic critical excitation methods for robust design. *Advances in Structural Engineering*, 8, 349–363.

Takewaki, I., Murakami, S., Fujita, K., Yoshitomi, S. & Tsuji, M. (2011) The 2011 off the Pacific coast of Tohoku earthquake and response of high-rise buildings under long-period ground motions. *Soil Dynamics and Earthquake Engineering*, 31, 1511–1528.

Tremblay, R. & Poncet, L. (2005) Seismic performance of concentrically braced steel frames in multi-story buildings with mass irregularity. *Journal of Structural Engineering*, 131, 1363–1375.

Trombetti, T. & Conte, J. (2005) New insight into and simplified approach to seismic analysis of torsionally coupled one-story, elastic systems. *Journal of Sound and Vibration*, 286, 265–312.

Trombetti, T., Silvestri, S., Gasparini, G., Pintucchi, B. & De Stefano, M. (2008) Numerical verification of the effectiveness of the "Alpha" method for the estimation of the maximum rotational elastic response of eccentric systems. *Journal of Earthquake Engineering*, 12, 249–280.

Truman, K.Z. & Cheng, F.Y. (1990) Optimum assessment of irregular three-dimensional seismic buildings. *Journal of Structural Engineering*, 116, 3324–3337.

Tso, W. & Bozorgnia, Y. (1986) Effective eccentricity for inelastic seismic response of buildings. *Earthquake Engineering & Structural Dynamics*, 14, 413–427.

Tso, W. & Myslimaj, B. (2003) A yield displacement distribution-based approach for strength assignment to lateral force-resisting elements having strength dependent stiffness. *Earthquake Engineering & Structural Dynamics*, 32, 2319–2351.

Tso, W. & Sadek, A. (1985) Inelastic seismic response of simple eccentric structures. *Earthquake Engineering & Structural Dynamics*, 13, 255–269.

Tso, W. & Zhu, T. (1992) Design of torsionally unbalanced structural systems based on code provisions I: Ductility demand. *Earthquake Engineering & Structural Dynamics*, 21, 609–627.

UBC (1997) Structural engineering design provisions. *International Conference of Building Officials, Whittier, California*.

Van Thuat, D. (2013) Story strength demands of irregular frame buildings under strong earthquakes. *The Structural Design of Tall and Special Buildings*, 22, 687–699.

Varadharajan, S., Sehgal, V. & Saini, B. (2012) Review of different structural irregularities in buildings. *Journal of Structural Engineering*, 39, 393–418.

Varum, H., Dumaru, R., Furtado, A., Barbosa, A.R., Gautam, D. & Rodrigues, H. (2018) Seismic performance of buildings in Nepal after the Gorkha earthquake. In *Impacts and insights of the Gorkha earthquake*. pp. 47–63, Elsevier, Amsterdam.

Wahalathantri, B.L., Thambiratnam, D., Chan, T. & Fawzia, S. (2011) A material model for flexural crack simulation in reinforced concrete elements using ABAQUS. *Proceedings of the First International Conference on Engineering, Designing and Developing the Built Environment for Sustainable Wellbeing, Queensland University of Technology, Queensland, Australia*, 260–264.

Wang, Y. (2018) *Damage Assessment in Asymmetric Buildings Using Vibration Techniques*. Queensland University of Technology, Queensland, Australia.

Wen, W.-P., Zhai, C.-H., Li, S., Chang, Z. & Xie, L.-L. (2014) Constant damage inelastic displacement ratios for the near-fault pulse-like ground motions. *Engineering Structures*, 59, 599–607.

Wikipedia contributors, 'Flatiron District', *Wikipedia, The Free Encyclopedia*, 17 December 2019, 10:20 UTC, <https://en.wikipedia.org/w/index.php?title=Flatiron_District&oldid=931164871>

Wilson, E. & Habibullah, A. (1998) *Sap 2000 Integrated Finite Element Analysis and Design of Structures Basic Analysis Reference Manual*. Computers and Structures, Berkeley.

Wu, C.L., Loh, C.-H. & Yang, Y. (2005) Shake table tests on gravity load collapse of low-ductility RC frames under near-fault earthquake excitation. *Proceedings of the Advances in Experimental Structural Engineering, Nagoya, Japan*, 725–732.

Wu, C.L., Kuo, W.W., Yang, Y.S., Hwang, S.J., Elwood, K.J., Loh, C.H. & Moehle, J.P. (2009) Collapse of a nonductile concrete frame: Shaking table tests. *Earthquake Engineering & Structural Dynamics*, 38, 205–224.

Xiao, J., Pham, T.L. & Ding, T. (2015) Shake table test on seismic response of a precast frame with recycled aggregate concrete. *Advances in Structural Engineering*, 18, 1517–1534.

Xin, D., Daniell, J.E. & Wenzel, F. (2018) State of the art of fragility analysis for major building types in China with implications for intensity-PGA relationships. *Natural Hazards and Earth System Sciences*, 1–34.

Xu, L., Pan, J., Leung, C. & Yin, W. (2018) Shaking table tests on precast reinforced concrete and engineered cementitious composite/reinforced concrete composite frames. *Advances in Structural Engineering*, 21, 824–837.

Ye, X., Ni, Y. & Yin, J. (2013) Safety monitoring of railway tunnel construction using FBG sensing technology. *Advances in Structural Engineering*, 16, 1401–1409.

Ye, X., Su, Y. & Han, J. (2014) Structural health monitoring of civil infrastructure using optical fiber sensing technology: A comprehensive review. *The Scientific World Journal*, 2014, 1–11.

Yenidogan, C., Yokoyama, R., Nagae, T., Tahara, K., Tosauchi, Y., Kajiwara, K. & Ghannoum, W. (2018) Shake table test of a full-scale four-story reinforced concrete structure and numerical representation of overall response with modified IMK model. *Bulletin of Earthquake Engineering*, 16, 2087–2118.

Yön, B., Sayın, E. & Onat, O. (2017) Earthquake and structural damages. *Earthquakes-Tectonics, Hazard and Risk Mitigation*, 319–339.

Zhang, C., Alam, Z. & Samali, B. (2016) Evaluating contradictory relationship between floor rotation and torsional irregularity coefficient under varying orientations of ground motion. *Earthquakes and Structures*, 11, 1027–1041.

Zhang, H., Kuang, J.S. & Yuen, T.Y. (2017) Low-seismic damage strategies for infilled RC frames: Shake-table tests. *Earthquake Engineering & Structural Dynamics*, 46, 2419–2438.

Zhao, B., Taucer, F. & Rossetto, T. (2009) Field investigation on the performance of building structures during the 12 May 2008 Wenchuan earthquake in China. *Engineering Structures*, 31, 1707–1723.

Zhou, Z., Huang, M., Huang, L., Ou, J. & Chen, G. (2011) An optical fiber Bragg grating sensing system for scour monitoring. *Advances in Structural Engineering*, 14, 67–78.

Zhu, T. & Tso, W. (1992) Design of torsionally unbalanced structural systems based on code provisions II: Strength distribution. *Earthquake Engineering & Structural Dynamics*, 21, 629–644.

Local response of RC model

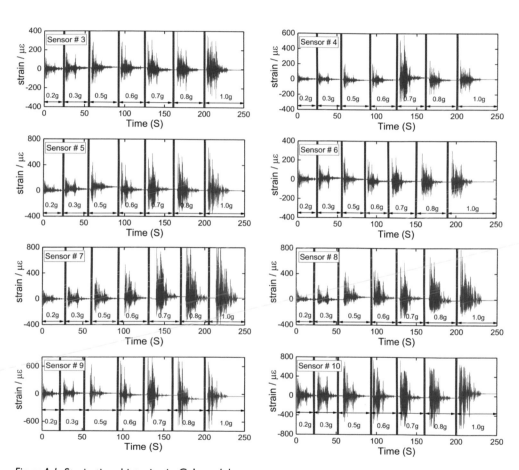

Figure A.1 Strain time histories in C-1 model

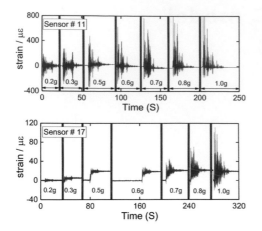
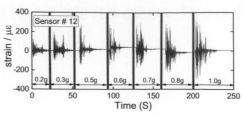

Figure A.1 (Continued)

Local response of steel models

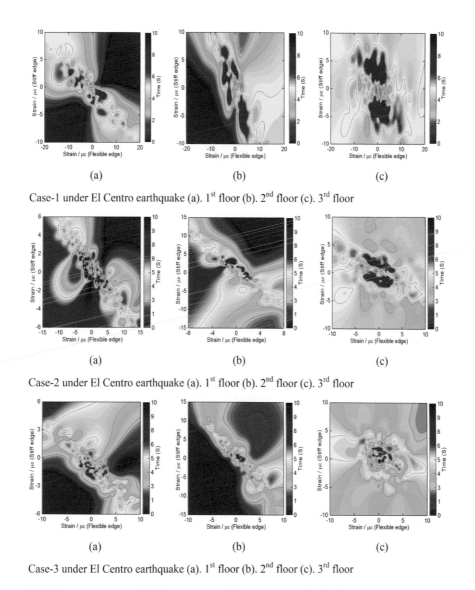

Case-1 under El Centro earthquake (a). 1st floor (b). 2nd floor (c). 3rd floor

Case-2 under El Centro earthquake (a). 1st floor (b). 2nd floor (c). 3rd floor

Case-3 under El Centro earthquake (a). 1st floor (b). 2nd floor (c). 3rd floor

Figure B.1 All asymmetric cases of bi-eccentric S-1 structure under El Centro earthquake

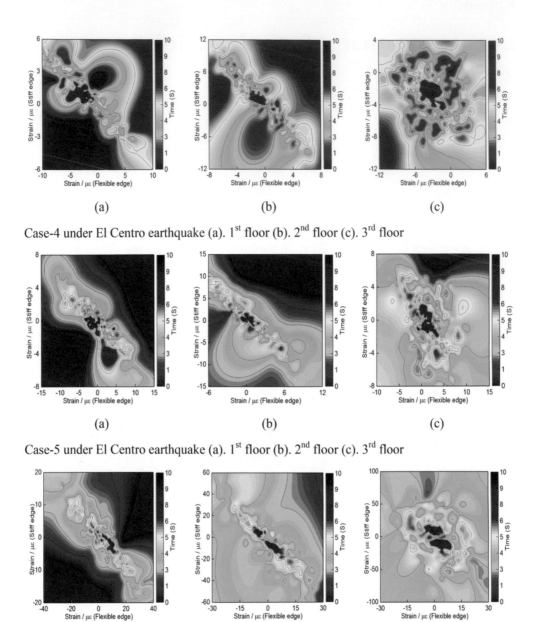

Case-4 under El Centro earthquake (a). 1st floor (b). 2nd floor (c). 3rd floor

Case-5 under El Centro earthquake (a). 1st floor (b). 2nd floor (c). 3rd floor

Case-6 under El Centro earthquake (a). 1st floor (b). 2nd floor (c). 3rd floor

Figure B.1 (Continued)

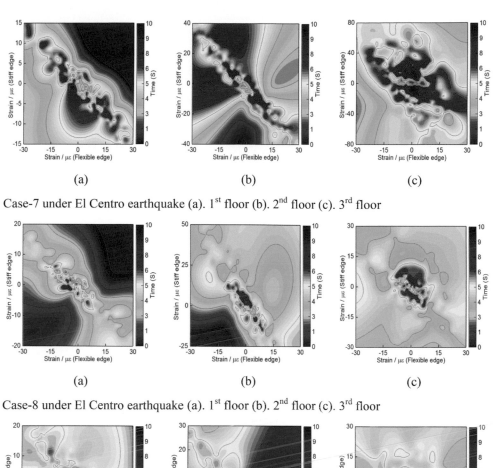

(a)   (b)   (c)

Case-7 under El Centro earthquake (a). 1st floor (b). 2nd floor (c). 3rd floor

(a)   (b)   (c)

Case-8 under El Centro earthquake (a). 1st floor (b). 2nd floor (c). 3rd floor

(a)   (b)   (c)

Case-9 under El Centro earthquake (a). 1st floor (b). 2nd floor (c). 3rd floor

Figure B.1 (Continued)

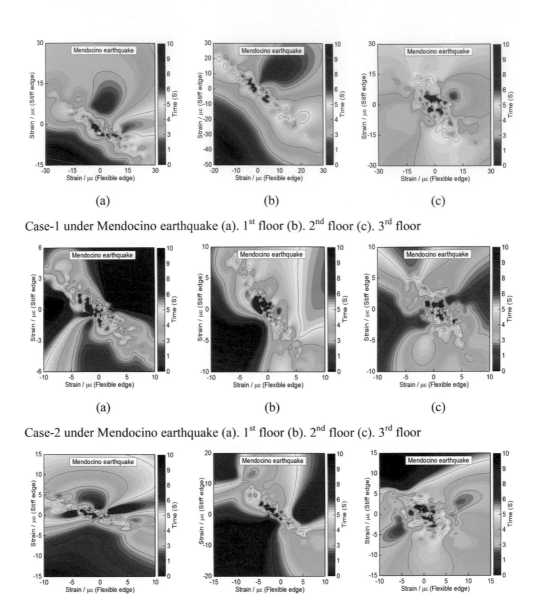

Case-1 under Mendocino earthquake (a). 1st floor (b). 2nd floor (c). 3rd floor

Case-2 under Mendocino earthquake (a). 1st floor (b). 2nd floor (c). 3rd floor

Case-3 under Mendocino earthquake (a). 1st floor (b). 2nd floor (c). 3rd floor

Figure B.2 All asymmetric cases of bi-eccentric S-1 structure under Mendocino earthquake

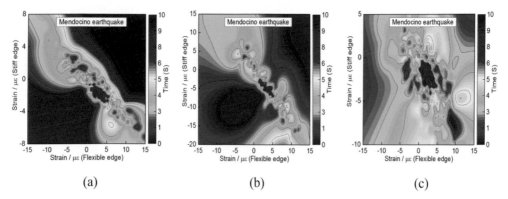

Case-4 under Mendocino earthquake (a). 1st floor (b). 2nd floor (c). 3rd floor

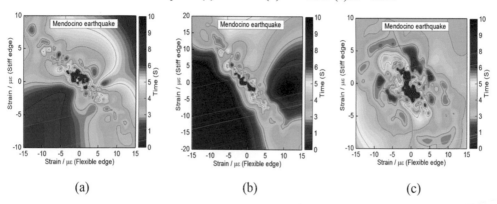

Case-5 under Mendocino earthquake (a). 1st floor (b). 2nd floor (c). 3rd floor

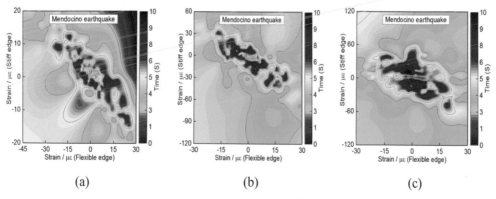

Case-6 under Mendocino earthquake (a). 1st floor (b). 2nd floor (c). 3rd floor

Figure B.2 (Continued)

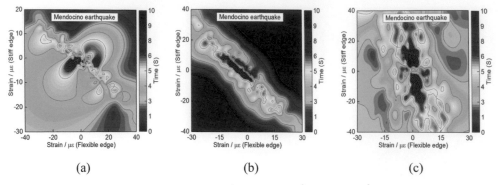

Case-7 under Mendocino earthquake (a). 1st floor (b). 2nd floor (c). 3rd floor

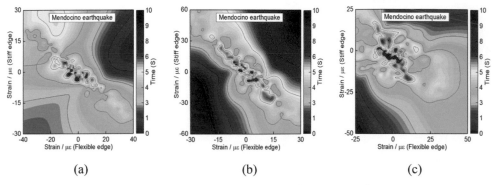

Case-8 under Mendocino earthquake (a). 1st floor (b). 2nd floor (c). 3rd floor

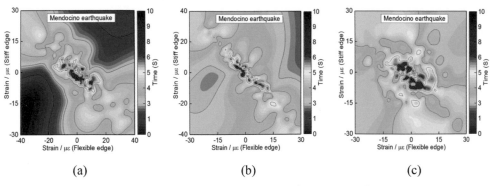

Case-9 under Mendocino earthquake (a). 1st floor (b). 2nd floor (c). 3rd floor

Figure B.2 (Continued)

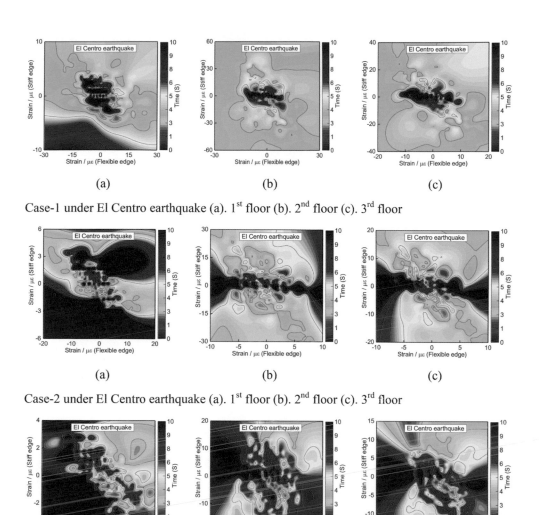

Case-1 under El Centro earthquake (a). 1st floor (b). 2nd floor (c). 3rd floor

Case-2 under El Centro earthquake (a). 1st floor (b). 2nd floor (c). 3rd floor

Case-3 under El Centro earthquake (a). 1st floor (b). 2nd floor (c). 3rd floor

Figure B.3 All asymmetric cases of mono-symmetric S-1 structure under El Centro earthquake

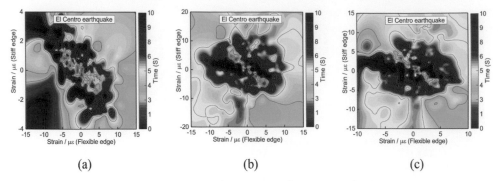

(a)   (b)   (c)

Case-4 under El Centro earthquake (a). 1st floor (b). 2nd floor (c). 3rd floor

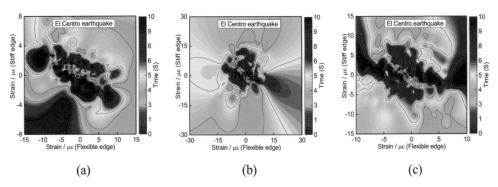

(a)   (b)   (c)

Case-5 under El Centro earthquake (a). 1st floor (b). 2nd floor (c). 3rd floor

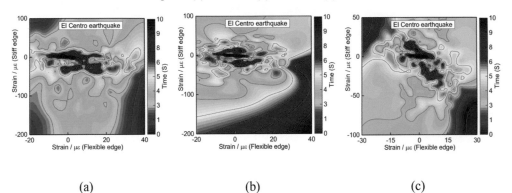

(a)   (b)   (c)

Case-6 under El Centro earthquake (a). 1st floor (b). 2nd floor (c). 3rd floor

Figure B.3 (Continued)

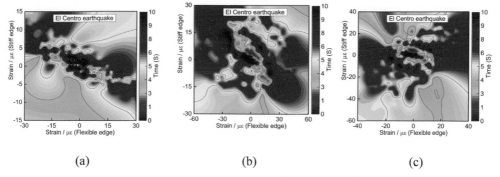

Case-7 under El Centro earthquake (a). 1st floor (b). 2nd floor (c). 3rd floor

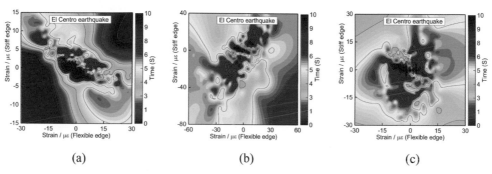

Case-8 under El Centro earthquake (a). 1st floor (b). 2nd floor (c). 3rd floor

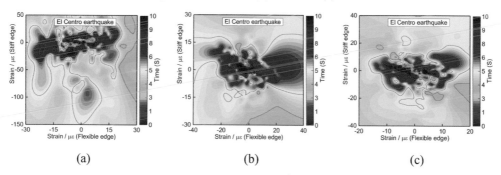

Case-9 under El Centro earthquake (a). 1st floor (b). 2nd floor (c). 3rd floor

Figure B.3 (Continued)

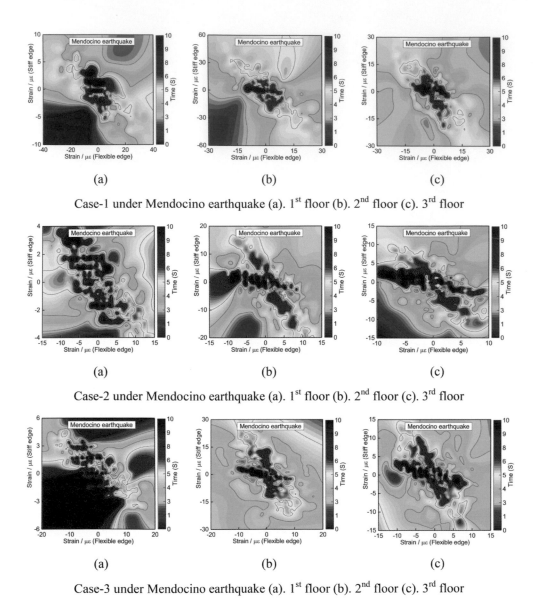

Case-1 under Mendocino earthquake (a). 1st floor (b). 2nd floor (c). 3rd floor

Case-2 under Mendocino earthquake (a). 1st floor (b). 2nd floor (c). 3rd floor

Case-3 under Mendocino earthquake (a). 1st floor (b). 2nd floor (c). 3rd floor

Figure B.4 All asymmetric cases of mono-symmetric S-1 structure under Mendocino earthquake

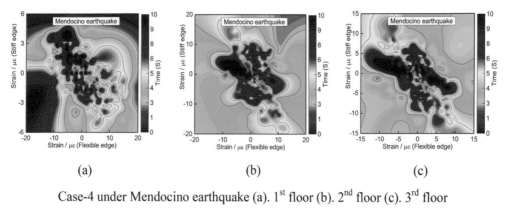

Case-4 under Mendocino earthquake (a). 1st floor (b). 2nd floor (c). 3rd floor

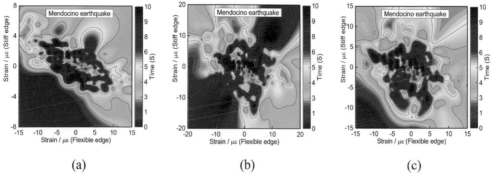

Case-5 under Mendocino earthquake (a). 1st floor (b). 2nd floor (c). 3rd floor

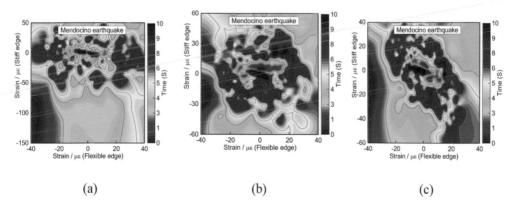

Case-6 under Mendocino earthquake (a). 1st floor (b). 2nd floor (c). 3rd floor

Figure B.4 (Continued)

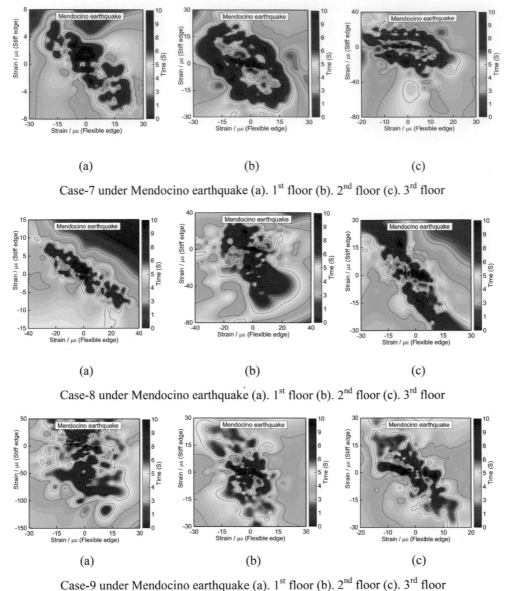

Case-7 under Mendocino earthquake (a). 1st floor (b). 2nd floor (c). 3rd floor

Case-8 under Mendocino earthquake (a). 1st floor (b). 2nd floor (c). 3rd floor

Case-9 under Mendocino earthquake (a). 1st floor (b). 2nd floor (c). 3rd floor

Figure B.4 (Continued)

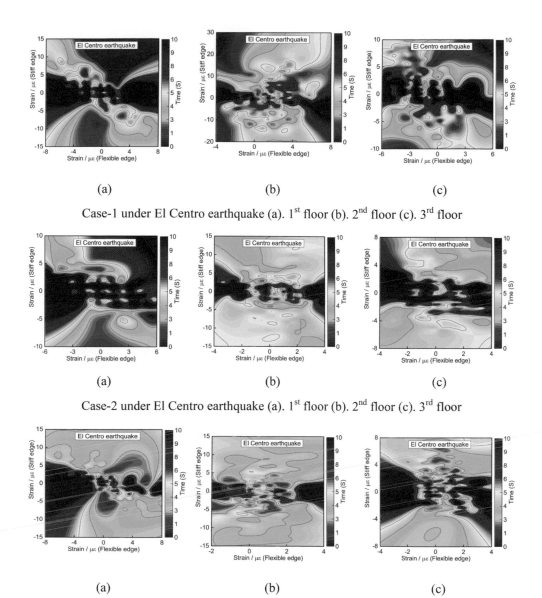

(a)   (b)   (c)

Case-1 under El Centro earthquake (a). 1st floor (b). 2nd floor (c). 3rd floor

(a)   (b)   (c)

Case-2 under El Centro earthquake (a). 1st floor (b). 2nd floor (c). 3rd floor

(a)   (b)   (c)

Case-3 under El Centro earthquake (a). 1st floor (b). 2nd floor (c). 3rd floor

Figure B.5 All asymmetric cases of S-2 structure under El Centro earthquake

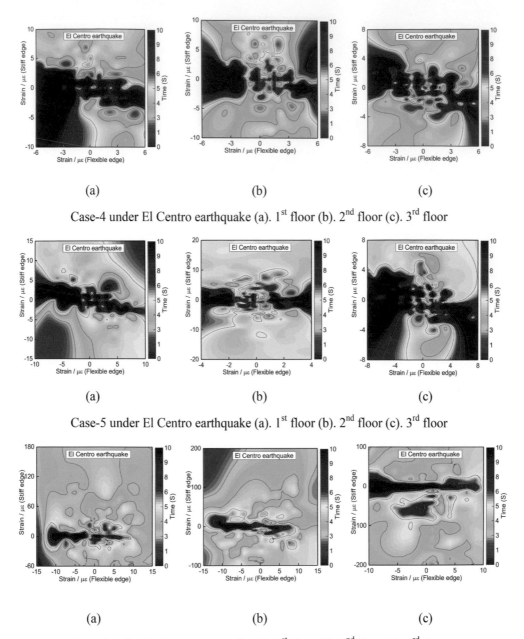

Case-4 under El Centro earthquake (a). 1st floor (b). 2nd floor (c). 3rd floor

Case-5 under El Centro earthquake (a). 1st floor (b). 2nd floor (c). 3rd floor

Case-6 under El Centro earthquake (a). 1st floor (b). 2nd floor (c). 3rd floor

Figure B.5 (Continued)

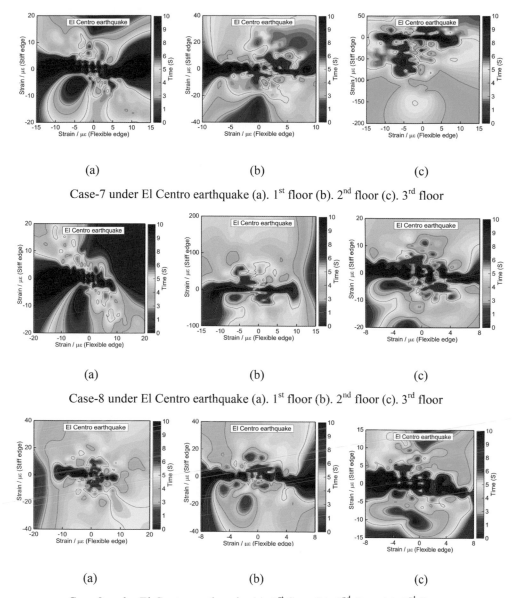

(a)   (b)   (c)

Case-7 under El Centro earthquake (a). 1st floor (b). 2nd floor (c). 3rd floor

(a)   (b)   (c)

Case-8 under El Centro earthquake (a). 1st floor (b). 2nd floor (c). 3rd floor

(a)   (b)   (c)

Case-9 under El Centro earthquake (a). 1st floor (b). 2nd floor (c). 3rd floor

Figure B.5 (Continued)

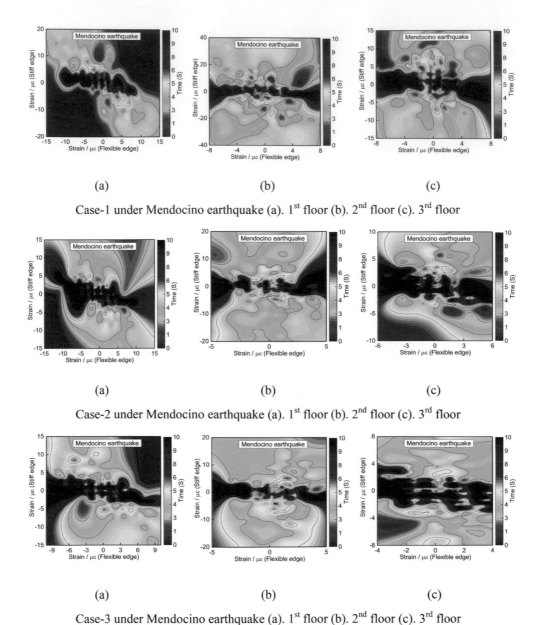

Case-1 under Mendocino earthquake (a). 1st floor (b). 2nd floor (c). 3rd floor

Case-2 under Mendocino earthquake (a). 1st floor (b). 2nd floor (c). 3rd floor

Case-3 under Mendocino earthquake (a). 1st floor (b). 2nd floor (c). 3rd floor

Figure B.6 All asymmetric cases of S-2 structure under Mendocino earthquake

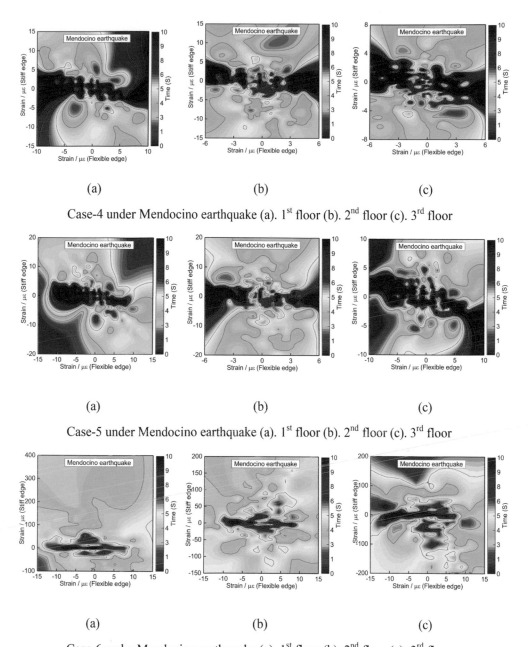

(a)   (b)   (c)

Case-4 under Mendocino earthquake (a). 1st floor (b). 2nd floor (c). 3rd floor

(a)   (b)   (c)

Case-5 under Mendocino earthquake (a). 1st floor (b). 2nd floor (c). 3rd floor

(a)   (b)   (c)

Case-6 under Mendocino earthquake (a). 1st floor (b). 2nd floor (c). 3rd floor

Figure B.6 (Continued)

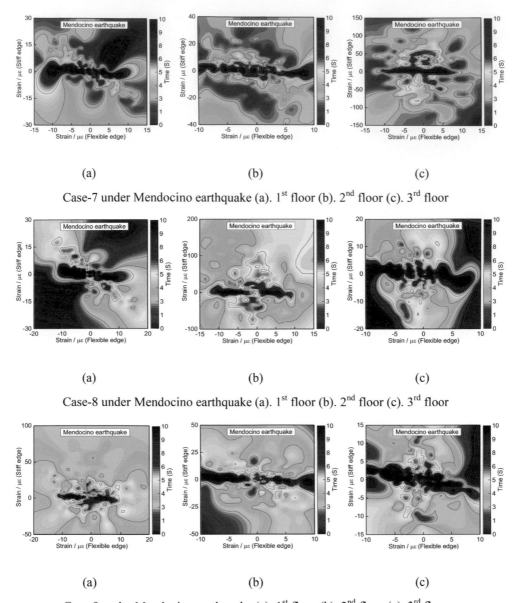

(a)   (b)   (c)

Case-7 under Mendocino earthquake (a). 1st floor (b). 2nd floor (c). 3rd floor

(a)   (b)   (c)

Case-8 under Mendocino earthquake (a). 1st floor (b). 2nd floor (c). 3rd floor

(a)   (b)   (c)

Case-9 under Mendocino earthquake (a). 1st floor (b). 2nd floor (c). 3rd floor

Figure B.6 (Continued)

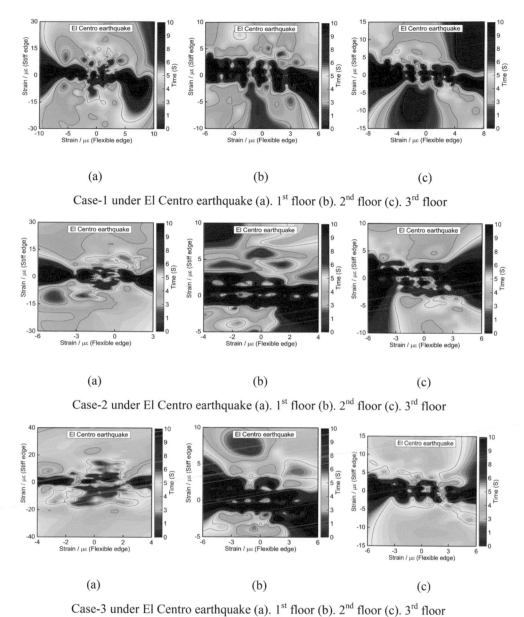

Case-1 under El Centro earthquake (a). 1st floor (b). 2nd floor (c). 3rd floor

Case-2 under El Centro earthquake (a). 1st floor (b). 2nd floor (c). 3rd floor

Case-3 under El Centro earthquake (a). 1st floor (b). 2nd floor (c). 3rd floor

Figure B.7 All asymmetric cases of S-3 structure under El Centro earthquake

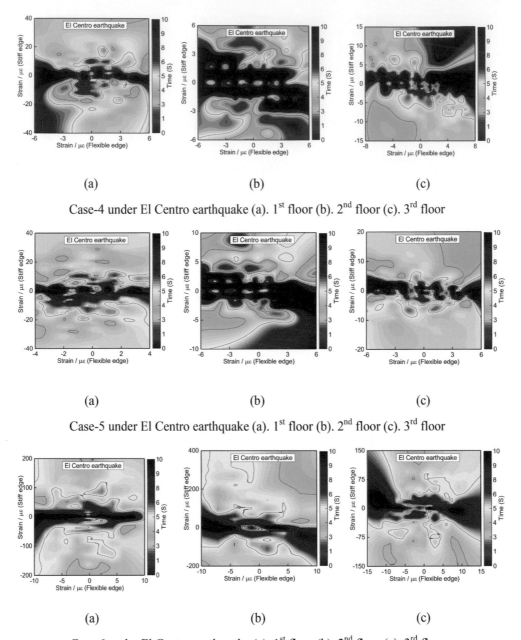

(a)   (b)   (c)

Case-4 under El Centro earthquake (a). 1st floor (b). 2nd floor (c). 3rd floor

(a)   (b)   (c)

Case-5 under El Centro earthquake (a). 1st floor (b). 2nd floor (c). 3rd floor

(a)   (b)   (c)

Case-6 under El Centro earthquake (a). 1st floor (b). 2nd floor (c). 3rd floor

Figure B.7 (Continued)

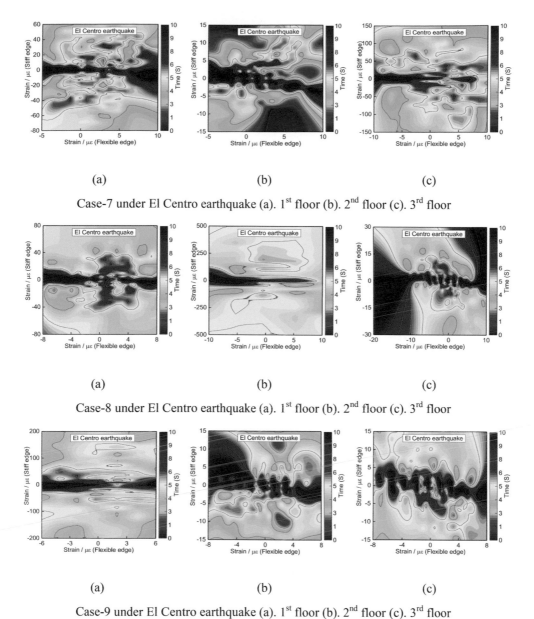

Case-7 under El Centro earthquake (a). 1st floor (b). 2nd floor (c). 3rd floor

Case-8 under El Centro earthquake (a). 1st floor (b). 2nd floor (c). 3rd floor

Case-9 under El Centro earthquake (a). 1st floor (b). 2nd floor (c). 3rd floor

Figure B.7 (Continued)

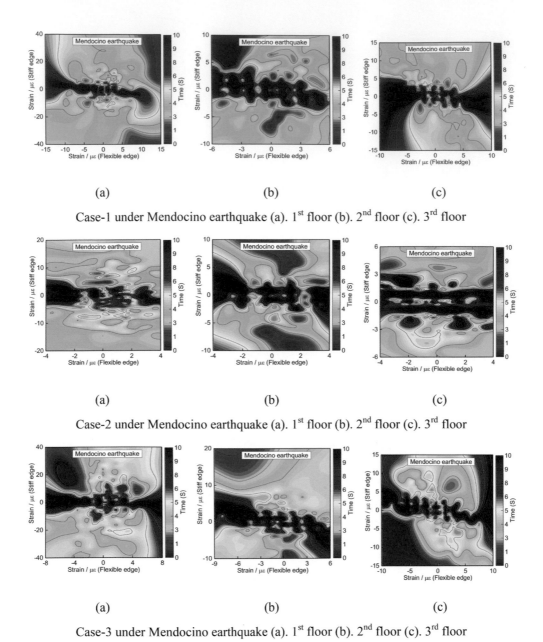

Case-1 under Mendocino earthquake (a). 1st floor (b). 2nd floor (c). 3rd floor

Case-2 under Mendocino earthquake (a). 1st floor (b). 2nd floor (c). 3rd floor

Case-3 under Mendocino earthquake (a). 1st floor (b). 2nd floor (c). 3rd floor

Figure B.8 All asymmetric cases of S-3 structure under Mendocino earthquake

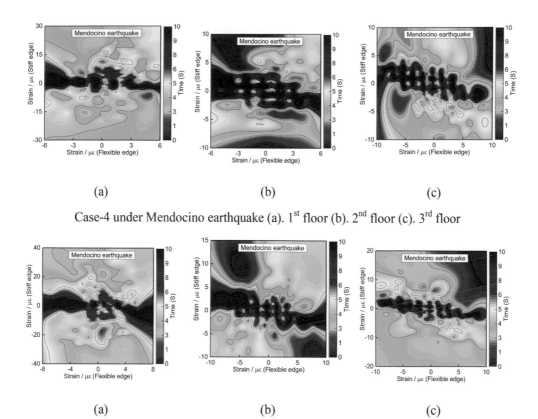

(a)   (b)   (c)

Case-4 under Mendocino earthquake (a). 1st floor (b). 2nd floor (c). 3rd floor

(a)   (b)   (c)

Case-5 under Mendocino earthquake (a). 1st floor (b). 2nd floor (c). 3rd floor

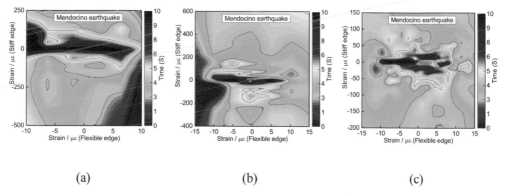

(a)   (b)   (c)

Case-6 under Mendocino earthquake (a). 1st floor (b). 2nd floor (c). 3rd floor

Figure B.8 (Continued)

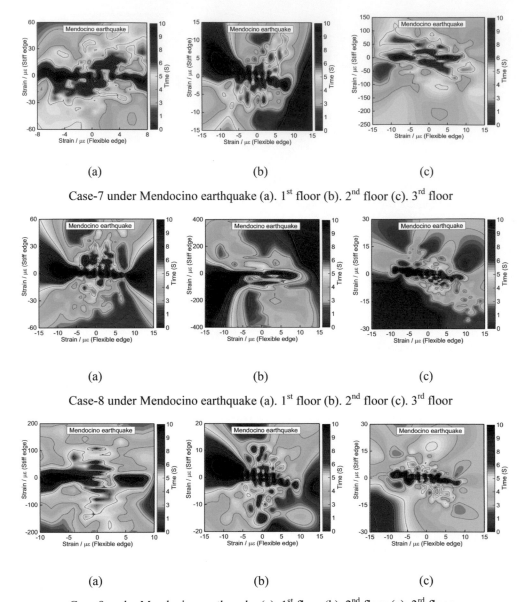

Case-7 under Mendocino earthquake (a). 1st floor (b). 2nd floor (c). 3rd floor

Case-8 under Mendocino earthquake (a). 1st floor (b). 2nd floor (c). 3rd floor

Case-9 under Mendocino earthquake (a). 1st floor (b). 2nd floor (c). 3rd floor

Figure B.8 (Continued)

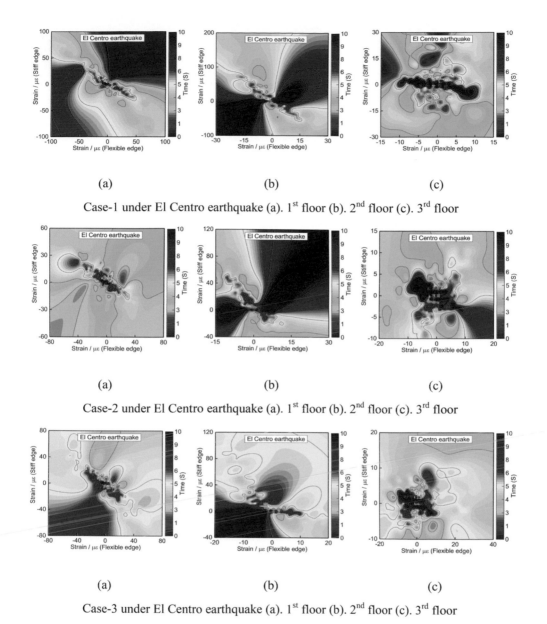

(a)   (b)   (c)

Case-1 under El Centro earthquake (a). 1st floor (b). 2nd floor (c). 3rd floor

(a)   (b)   (c)

Case-2 under El Centro earthquake (a). 1st floor (b). 2nd floor (c). 3rd floor

(a)   (b)   (c)

Case-3 under El Centro earthquake (a). 1st floor (b). 2nd floor (c). 3rd floor

Figure B.9 All asymmetric cases of S-4 structure under El Centro earthquake

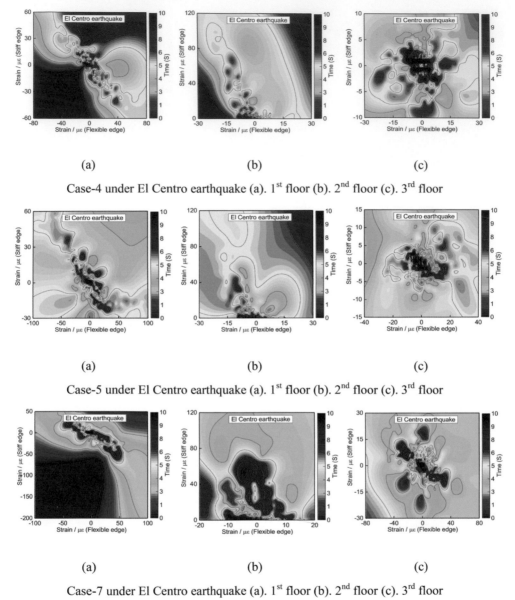

(a)   (b)   (c)

Case-4 under El Centro earthquake (a). 1st floor (b). 2nd floor (c). 3rd floor

(a)   (b)   (c)

Case-5 under El Centro earthquake (a). 1st floor (b). 2nd floor (c). 3rd floor

(a)   (b)   (c)

Case-7 under El Centro earthquake (a). 1st floor (b). 2nd floor (c). 3rd floor

Figure B.9 (Continued)

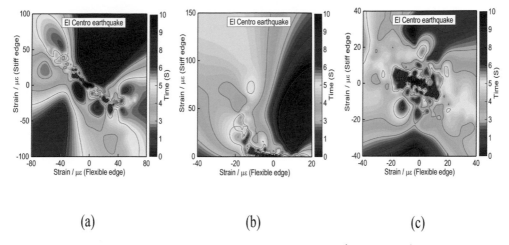

(a)   (b)   (c)

Case-8 under El Centro earthquake (a). 1st floor (b). 2nd floor (c). 3rd floor

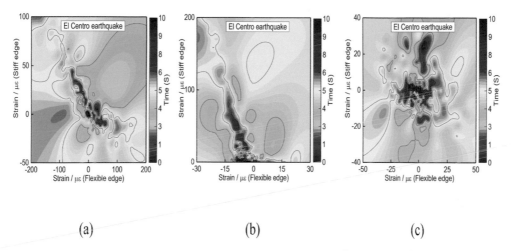

(a)   (b)   (c)

Case-9 under El Centro earthquake (a). 1st floor (b). 2nd floor (c). 3rd floor

Figure B.9 (Continued)

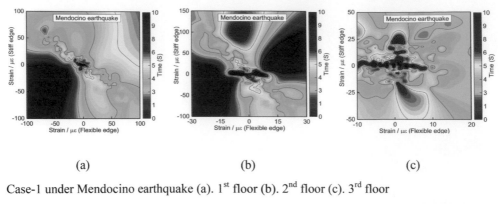

(a)   (b)   (c)

Case-1 under Mendocino earthquake (a). 1st floor (b). 2nd floor (c). 3rd floor

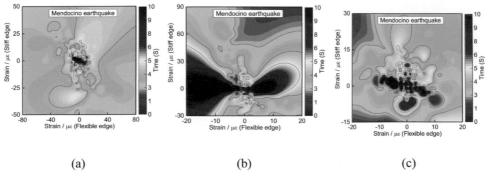

(a)   (b)   (c)

Case-2 under Mendocino earthquake (a). 1st floor (b). 2nd floor (c). 3rd floor

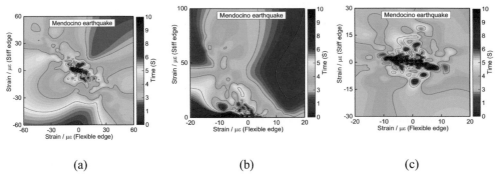

(a)   (b)   (c)

Case-3 under Mendocino earthquake (a). 1st floor (b). 2nd floor (c). 3rd floor

Figure B.10 All asymmetric cases of S-4 structure under Mendocino earthquake

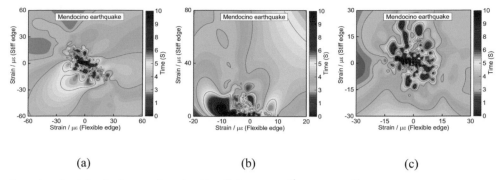

(a)   (b)   (c)

Case-4 under Mendocino earthquake (a). 1st floor (b). 2nd floor (c). 3rd floor

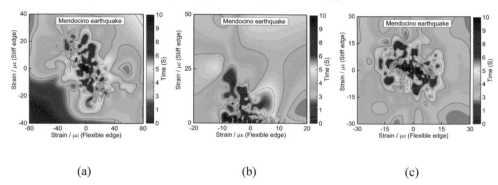

(a)   (b)   (c)

Case-5 under Mendocino earthquake (a). 1st floor (b). 2nd floor (c). 3rd floor

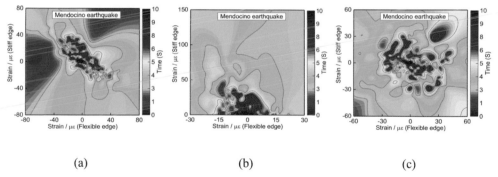

(a)   (b)   (c)

Case-7 under Mendocino earthquake (a). 1st floor (b). 2nd floor (c). 3rd floor

Figure B.10 (Continued)

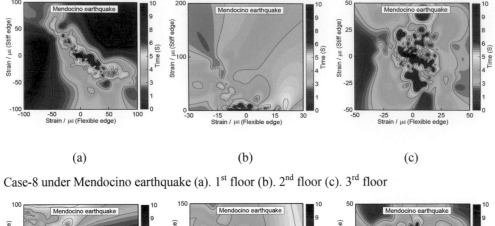

(a)	(b)	(c)

Case-8 under Mendocino earthquake (a). 1st floor (b). 2nd floor (c). 3rd floor

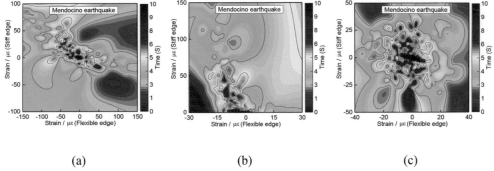

(a)	(b)	(c)

Case-9 under Mendocino earthquake (a). 1st floor (b). 2nd floor (c). 3rd floor

Figure B.10 (Continued)

Global behavior of steel models

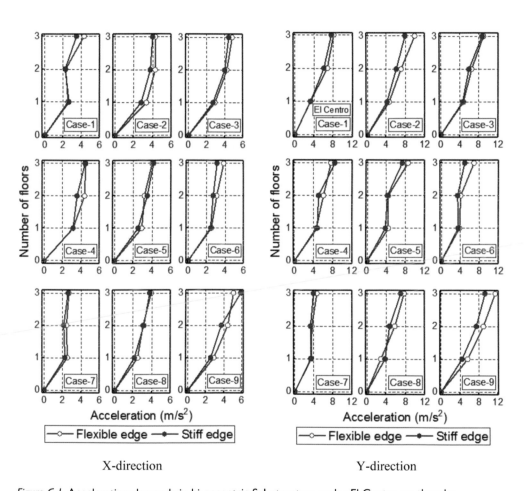

X-direction   Y-direction

Figure C.1 Acceleration demands in bi-eccentric S-1 structure under El Centro earthquake

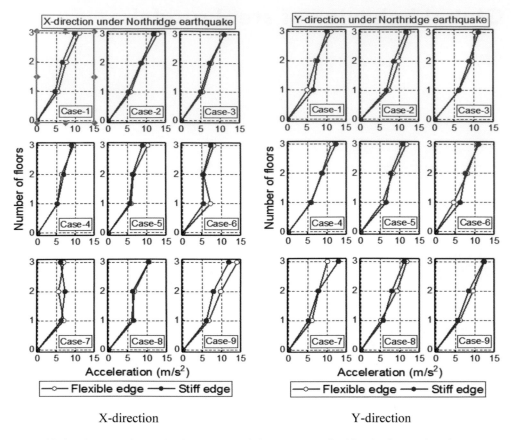

X-direction   Y-direction

Figure C.2 Acceleration demands of bi-eccentric S-1 structure under Northridge earthquake

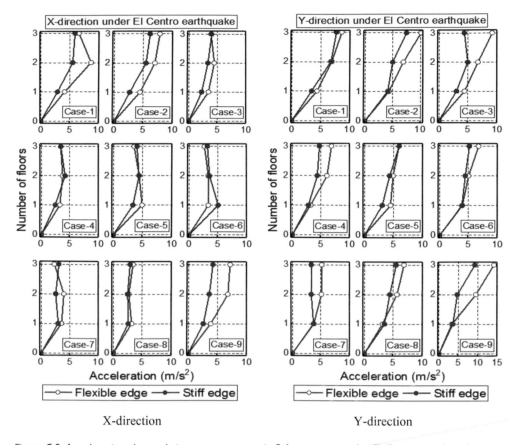

Figure C.3 Acceleration demands in mono-symmetric S-1 structure under El Centro earthquake

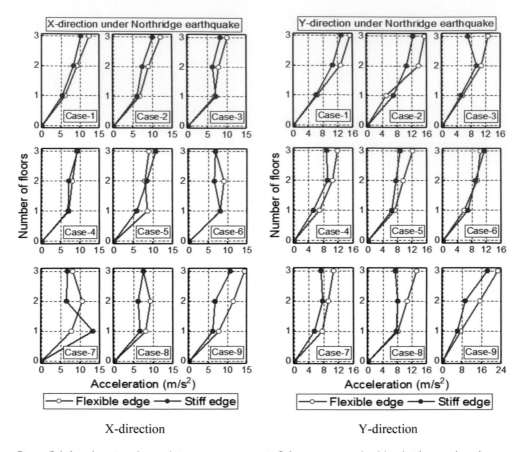

X-direction   Y-direction

Figure C.4 Acceleration demands in mono-symmetric S-1 structure under Northridge earthquake

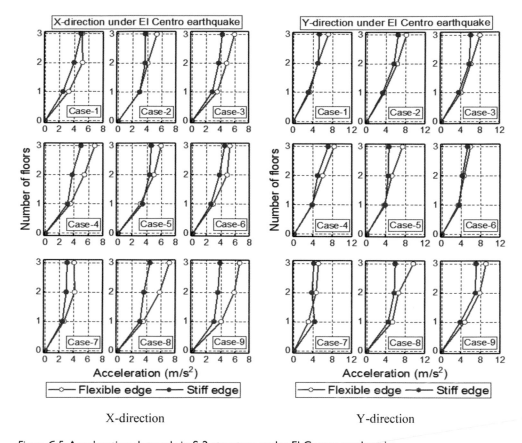

Figure C.5 Acceleration demands in S-2 structure under El Centro earthquake

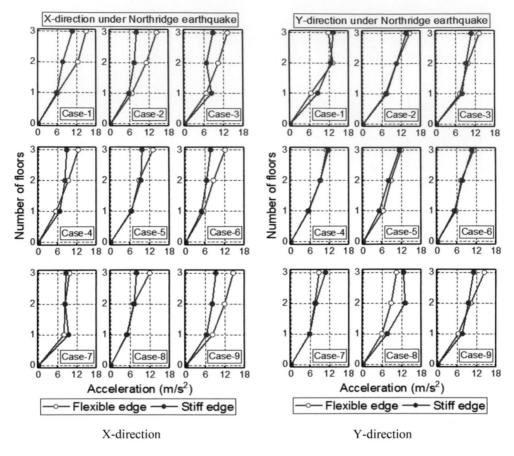

X-direction   Y-direction

Figure C.6 Acceleration demands in S-2 structure under Northridge earthquake

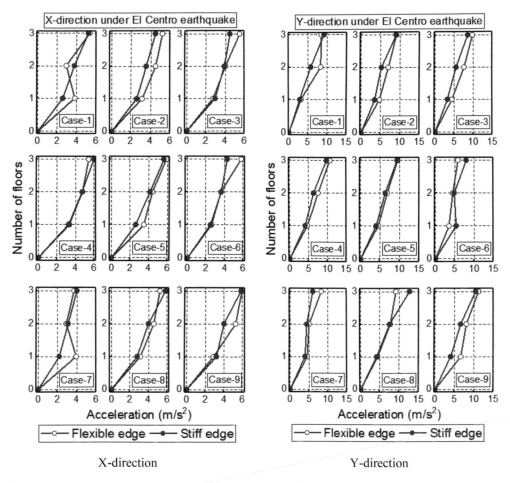

X-direction   Y-direction

Figure C.7 Acceleration demands in S-3 structure under El Centro earthquake

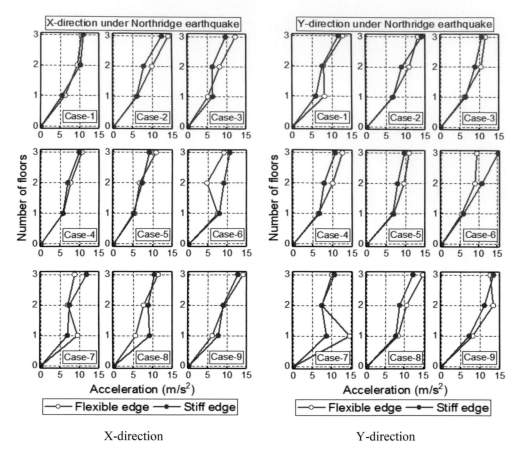

X-direction   Y-direction

Figure C.8 Acceleration demands in S-3 structure under Northridge earthquake

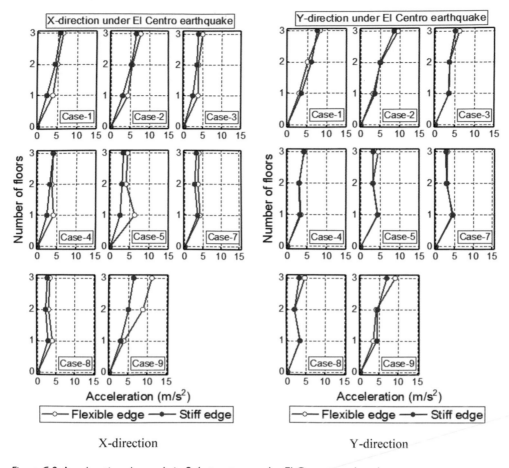

Figure C.9 Acceleration demands in S-4 structure under El Centro earthquake

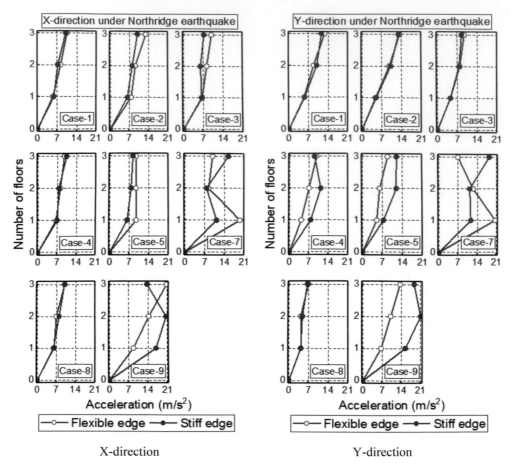

Figure C.10 Acceleration demands in S-4 structure under Northridge earthquake

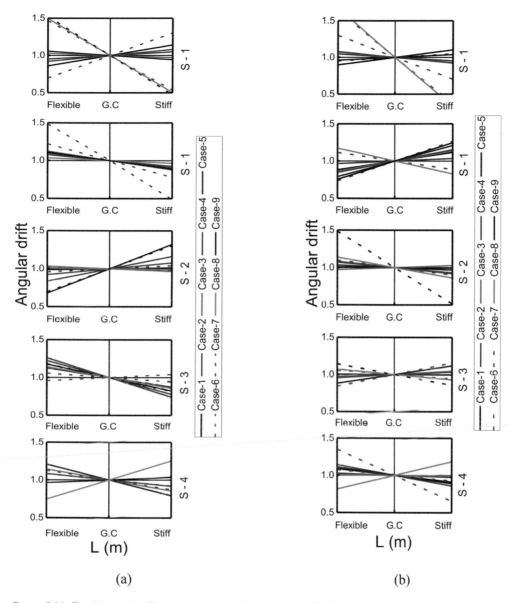

Figure C.11 Flexible and stiff rotation in steel models: (a) El Centro earthquake (b) Northridge earthquake